THE GULF STREAM

By the same author

World Beneath the Oceans
Physics of the Earth
Under the Deep Oceans
North Sea Oil: the Great Gamble
Using the Oceans (editor)
The Earth's Mantle (editor)

THE JOHN DAY COMPANY • NEW YORK
AN Intext PUBLISHER

THE GULF STREAM
T.F. GASKELL

First American Edition 1973

The John Day Company, 257 Park Avenue South, New York, N.Y. 10010

Library of Congress Cataloging in Publication Data

Gaskell, Thomas Frohock.
The Gulf Stream.

Bibliography: p.
1. Gulf Stream. I. Title.
GC296.G9G3 1973 551.4'71 72-8299
ISBN: 0-381-98213-0

Printed in the United States of America

To my oceanographic friends who continue to work in all weathers to discover how the earth's waters circulate

CONTENTS

ILLUSTRATIONS

MAPS

FOREWORD

This book was started several years ago, and is the brainchild of Kenneth Parker of Cassell, who decided that there was a need for a popular account of the Gulf Stream, including aspects 'historical, geographical, navigational, oceanographical, meteorological, political, naval and military, hydrographical, biological, piscatorial . . . a biography of a phenomenon which could cover its past, present and future!' The book has taken a long time to complete because a great deal of looking into old publications is needed to compile a work of this type, and as a part-time author it is difficult to fit library research into the week-ends. As a subject for the general reader it must also of necessity leave out all the elegant and mystifying mathematics with which most books on the subject can be partly filled, but which convey something to only a very few experts.

The more one thinks about the Gulf Stream the more it becomes apparent that this is a feature of our earth with a two-fold meaning. In the first instance there is the observable effect of a fast-moving body of water traversing the North Atlantic, an effect which has been of importance to navigators and inventors in the past, while secondly there is the more nebulous fact that the Gulf Stream is only part of a whole, a general circulation of water in the ocean, which is inexorably determined by the rotation of the earth together with its atmosphere. Just as the river-like motion of the Gulf Stream has its practical effect on our life, so the imperfectly understood broader aspects of the Gulf Stream are of great concern in the long-term future of the earth. The odd assortment of subjects relating to the Gulf Stream that are covered in this book will, it is hoped, provide starting points for many lines of thought to those wishing to plan a decent world. It is hoped that politicians and others who control the financing of research will realize the importance of future well-planned oceanographic experiments designed to verify the different hypotheses of how the ocean circulation takes place and what interaction there is between air and water.

It would not have been possible to attempt this book without the blessing of Dr. Henry Stommel and without the use of his classical book on the subject—*The Gulf Stream.* Dr. Stommel is still the leader of many of the pioneering ideas in the study of movement of the waters of the sea. Dr. G. Wüst was so kind as to send his original papers on early calculations based on his own measurements and on the direct current observations of Lieutenant Pillsbury of the U.S. Coast and Geodetic Survey. Dr. John Swallow, an oceanographic partner for many years, has provided not only information about the deep ocean direct current movements, but also corrective advice concerning the theories of ocean circulation. One of the best textbooks, and definitely one for further reading, is *The Waters of the Sea* by Dr. P. Groen, and my thanks go to him for providing such a helpful book of reference. Dr. Jacques Piccard kindly checked the chapter on the Gulf Stream Project, while Dr. Anthony Michaelis provided many useful press cuttings together with ideas including library research into foreign books. Dr. Robert Dietz and Dr. Charles Bates, well known in U.S. marine geology and oceanography, have been kind enough to collect some of the Americana, and Mr. Wolsey of Northern Ireland made an especial effort to gather together the details of the eel for me. Dr. Arthur Bourne, who heads O.R.C.A. (Ocean Resources Conservation Association), as always, has been a great help with suggestions of lines that should be followed. I would like to extend my thanks to the Librarian of the Royal Geographical Society and to the members of BP's Technical Information and Library Service who have given me great assistance in finding material. There are many books as well as those mentioned that have been of value to me, and these are listed in a Bibliography on p. 162.

It is impossible to write a book without the help of someone to type out the manuscript and this Miss Lynne Jenkins has done most expeditiously.

Week-end writing is a trial for the family and mine have borne the strain without complaint. I would like to express here my deepest thanks for the encouragement and the peaceful conditions provided by my wife Joyce.

THE GULF STREAM

1

THE GULF STREAM IN HISTORY

There is an odd flow of water in the Atlantic Ocean which has had a profound effect on human beings. This current of water has been called a 'crazy river' by some; to others it is an understandable natural phenomenon. In the past some philosophers have wanted to change the flow of this current, while others have regarded it as immortal. Perhaps the sensible ones realize that Nature works so slowly that the length of man's life permits only a snapshot of the earth to be viewed. The changes that have taken place in the thousands of millions of years of geological history can only be deduced from the traces left from the past and from observations of the mechanisms of natural phenomena that are operating today.

The Gulf Stream has always affected life on earth. In the past it controlled the discovery of the Americas, and has been partly responsible for some of the more sordid episodes of history in places such as the Caribbean and South America between the fifteenth and eighteenth centuries. The Gulf Stream could control battles in the days of sail, and it can affect the tactics of submarine warfare today. Politics may well figure in future Gulf Stream history. Already the mineral wealth of the ocean floor is being parcelled out amongst the nations of the world. One day the warmth of the ocean currents may prove to be a cause for argument. Engineers, with the help of great earth-moving machines and atomic bombs, can alter the face of the earth. Proposals have been made to block the Straits of Gibraltar, or to cut the Isthmus of Panama or to close the Bering Straits. These large-scale alterations to the earth's surface may have far-reaching effects, some of which could be unpleasant and perhaps irreversible. In the early part of this century Mr. Carroll Livingston Riker of New York produced a scheme for controlling the Gulf Stream by forming a 200-mile jetty east from Newfoundland. Although this most elegant scheme, especially the ingenious way of achieving the inventor's aim, might commend itself to many, the repercussions would probably be great, both in time and distance. It is essential, therefore, that the way Nature operates such massive natural phenomena as the Gulf Stream should be studied with great care before human beings are allowed to tamper with the works.

The Gulf Stream, part fast-flowing and fairly well defined, part a mere drift of a vast body of water, may well be called a crazy river, but it is no crazier than the history that it has affected. The earliest known Greek geographical drawings have hints of a sea of weed which could be the Sargasso Sea, and hence an indication of voyaging into the Atlantic with the aid of the Canaries Current and the North Equatorial Drift in 150 B.C. The Phoenicians were probably familiar with the eastern parts of the Atlantic circulation, since they made voyages to Britain. This knowledge was no doubt passed on to the Arabs who, over a thousand years later, taught the Portuguese. Andrea Bianca's Atlantic map of 1436 shows a 'sea of berries' where the centre of the Atlantic current system should be. After defeating the Carthaginians, the heavy Roman hand, supported by foot-soldiers rather than sailors, probably stopped any early development of Atlantic exploration. The only experience of the Gulf Stream until the fifteenth century was possibly during the voyages of the Norsemen around A.D. 1000. Recent evidence suggests that the Vikings discovered 'Vinland', which could be the North American continent. Whether or not this is so, the Gulf Stream had played its part in producing a milder climate than exists today. Settlements were established in Greenland, and the long journeys from Europe could well have been extended to Labrador and Newfoundland. The Gulf Stream may well have been known for its assistance on the journeys home. There is a report from the bishopric of Skalholt in Iceland in 1347 that a battered vessel arrived from across the Atlantic. The craft was a small Greenland vessel which had lost its anchor and sails and had been swept eastwards by the Gulf Stream and its tributaries.

Bianca's 1436 map showed that voyages were being made far into the Atlantic during the fifteenth century. There is some evidence that the Portuguese were visiting the rich fishing-grounds off Newfoundland. Possibly some of the British and French cod fishermen pressed onwards from their Icelandic banks towards the shores of Canada. If they did so, they must have experienced the fogs that occur where the Gulf Stream meets the cold Labrador Current. Towards the end of the fifteenth century, when Columbus was planning his notable crossing of the Atlantic, it was known that there must be land to the West because 'sea beans' (the seed of a West Indian plant), bamboo stems and coconuts were occasionally washed up on European shores. These bits of flotsam indicated both a source to the west and a current moving eastwards.

In 1492 Columbus discovered America; at least according to history books, although in fact he was searching for the westward route to the East Indies. It is said that Columbus was inspired partly by a story he had read of a voyage made by the Queen of Sheba to Japan by a westward round-the-world route. It was during this voyage that Columbus, on 19 September, became aware of a westerly drift, the Canaries Current. This stream gave his ship an extra forty miles a day and probably ensured the success of his voyage. However, although Columbus on subsequent voyages noted that this helpful Atlantic current flowed into the Caribbean Sea between the Windward Islands, and left the Caribbean through the Yucatan Channel, he never realized that the Stream turned right about and provided an easy ride homewards.

John and Sebastian Cabot sailed up the western side of the Atlantic as far as Labrador in 1497, and observed the counter-current which runs in a south-westerly direction between the Gulf Stream and the American coast. Sebastian Cabot made the interesting observation that the beer in the hold of his ship fermented and turned sour because of unaccountable warmth below decks. (This was due, of course, to the high temperature of the Gulf Stream.) Apparently neither Columbus nor the Cabots connected the high temperature with the current. Columbus did, however, give the first authentic description of the Sargasso Sea, the idle hub of the North Atlantic circulating current system. He reported a seaweed like grass floating in bushels around the ship, and hauled a bunch of it on board. Some crabs were found in the weed, and the sea was warm enough to tempt the sailors to undress and go swimming. Later exaggerations depicted the Sargasso Sea as a solid mass of weed which trapped ships and held them fast to rot.

Sebastian de Ocampo sailed round Cuba in 1509 and must

have had to struggle against the Gulf Stream. However, no one reported the current, although it is so fast in that part of the world that it cannot be missed. Perhaps the navigators of those days were jealous of their knowledge and kept useful information to themselves and to their close friends. Perhaps they could not write very well. Whatever the reason, it was not until 1513 that Juan Ponce de León described the Florida Current, which is the fast-moving beginning of the Gulf Stream. Ponce de León sailed from Puerto Rico across the Stream towards Cape Canaveral (now Cape Kennedy) and then went south to the Tortugas.

Peter Martyr, a sixteenth-century Italian, and author of *De Rebus Oceanis et Novo Orbe* (1516), was the first person to give sensible thought to the Atlantic currents. He realized that the westward flowing North Equatorial Current must do one of three things. It could pile up large masses of water at the Brazilian coast, but the explorers of this area had never noticed such a phenomenon. There could be a passage into the Pacific, so that the Stream could continue round the world and back into the Atlantic, but the evidence from exploration was in favour of a continuous land barrier. This led Martyr to the third conclusion—that the westward Atlantic current was deflected by the land and diverted through the Florida Straits and up the coast of North America.

The next few years covered a period of intense activity in the Caribbean area, and by 1519 the Spanish ships had discovered the trick of sailing with favourable currents all the way—across the Atlantic to America with the Equatorial Current, and returning home through the Florida Straits, along the Gulf Stream as far as Cape Hatteras and then eastward to Spain. There does not seem to have been much written about the Gulf Stream, probably because the routing of Spanish treasure ships was quite reasonably kept secret on account of the growing menace of pirates.

In 1575 André Thevet maintained that the flow of water through the Florida Straits was the natural escape of the rivers such as the Mississippi, which fed the Gulf of Mexico. It is now known that this cannot possibly be true, since there is a discrepancy of more than a thousand to one in the volume of the flow of the Gulf Stream and that of the Mississippi. Another interesting reference concerns the temperature of the surface water in the Gulf Stream. Lescarbot wrote in 1609 that he had been sailing a few hundred miles east of the Newfoundland Banks in 1606 on a day when the air was cold, yet the water was very warm. Three days later, although it was 21 June, the ship suddenly ran into fog, and the sea was extremely cold too.

While the Spaniards were wiping out the inhabitants of the Caribbean area, and in turn were being attacked both by pirates and by the French, Dutch and British navies, the true explorers turned their activities to the north. Since there was no passage to the Pacific through the isthmus joining the Americas, a northern route was sought. The cold Labrador Current was crossed many times by persistent searchers such as Sir Humphrey Gilbert and Martin Frobisher. The branch of the Gulf Stream that runs north-east to Norway was observed, and Frobisher saw that it must be part of a general circulation. He did not, however, realize that the main Atlantic current system runs southward when it approaches Europe to give the wheel-like movement centred on the Sargasso Sea; but he did at least produce a consistent scheme, by running his return current round the north of Norway and thence by some unexplained route to the east coast of Africa. It was Isaac Vossius in 1663 who first postulated the ocean-wide clockwise motion that operates in the North Atlantic. The mechanism that made the water move was supposed to be a piling up of the sea at the Equator by the sun. When the mountain of water which followed the sun reached the American coast it was diverted north. This theory was soon followed by the first chart showing the Gulf Stream by Athanasius Kircher in 1665. As with many old charts, the facts are pleasantly interwoven with mythological figures and extraordinary phenomena. To those who prefer strictly factual material on their maps this may be irritating, but it does make the pictorial effect more interesting if unknown areas are filled with imaginative sketches rather than a blank space. Today the oil industry is picking up the threads of the old seekers of the North West Passage and may one day make a new ocean highway from eastern U.S.A. to Alaska.

Benjamin Franklin, that extraordinarily capable natural philosopher, started to put the Gulf Stream literally on the map when he collected the vast experience of the New England whaling skippers to produce his well known chart. Franklin was Postmaster-General in 1770 and was justifiably concerned because the mail packets took two weeks longer to cross the Atlantic than did merchant ships. A Nantucket sea captain, Timothy Folger, explained why this was so:

> We are well acquainted with the Stream because in our pursuit of whales, which keep to the sides of it but are not met within it, we run along the side and frequently cross it to change our side, and in crossing it have sometimes met and spoke with those packets who were in the middle of it and stemming it. We have informed them that they were stemming a current that was against them to the value of three miles an hour and advised them to cross it, but they were too wise to be councelled by simple American fishermen!

Benjamin Franklin wrote to the Secretary of the British Post Office:

> Craven Street, October 29, 1769
>
> Sir: Discoursing with Captain Folger . . . I received from him the following information, viz.: . . . that the whales are found generally near the edges of the Gulph* Stream, a strong current so called, which comes out of the Gulph of Florida, passing north-easterly along the coast of America, and then turning off most easterly, running at the rate of 4, 3–1/2, 3 and 2–1/2 miles an hour; that (people concerned in the whale fishery) . . . cruise along the edges of the stream in quest of whales . .; that they have opportunities of discovering the strength of it when their boats are out in pursuit of this fish, and happen to get into the stream while the ship is out of it, or out of the stream while the ship is in it, for then they are separated very fast, and would soon lose sight

*Franklin obviously took this odd spelling from the classical Greek. The 1911 *Encyclopaedia Britannica* states: 'The word "gulf", a portion of the sea enclosed by the coastline and usually taken as referring to a tract of water larger than a bay and smaller than a sea is derived through the French *golfe*, from the late Greek κόλφος, Classical Greek κόλπος, bosom, hence bay, or Latin *sinus*.

> of each other if care were not taken; that . . . they frequently . . . speak with ships bound from England to New York, Virginia, etc. . . . and it is supposed that their fear of Cape Sable shoals, George's Banks, or Nantucket shoals, hath induced them to keep so far to the southward as unavoidably to engage them in the same Gulph Stream, which occasions the length of their voyages, since . . . the current being 60 or 70 miles a day, is so much subtracted from the way they make through the water.
>
> At my request Captain Folger hath been so obliging as to mark for me on a chart the dimensions, course, and swiftness of the stream from its first coming out of the Gulph . . . ; and to give me withal some written directions whereby ships bound from the Banks of Newfoundland to New York may avoid the said stream, and yet be free of danger from the banks and shoals above mentioned. . . . With much esteem, I am, etc . . .
>
> Benjamin Franklin

Franklin attributed the Gulf Stream to a piling up of water at the American coast by the force of the prevailing trade winds. It is, of course, the winds that provide the driving force for the whole North Atlantic circulation, and any variations in level are caused by the winds. However, the circulation must be considered as a whole, since it does appear from recent work that there are places where currents apparently flow uphill (a phenomenon sometimes reported by yachtsmen). Franklin was one of the pioneers in using temperature to define the boundaries of the Gulf Stream. In addition to making surface measurements with a thermometer, to indicate whether the ship was in or out of the current, he attempted measurements down to 100 feet below the surface, first with a sample bottle, and later with a cask fitted with a valve at each end. The temperature method of tracking the currents led Captain Strickland at the turn of the century to discover the north-easterly area of the Gulf Stream which extends to Britain and Norway. Although this was probably known to the Norsemen, it had until this time escaped the attention of the chart-makers.

One of the early oceanographers, whose name has been given to one of the larger currents in the Pacific, was Alexander von Humboldt. During a stay in Havana he took the opportunity to study the Gulf Stream and he claims to have covered 25,000 miles sailing up and down its warm waters. An interesting calculation he made gave the time for a complete circuit of the North Atlantic current system as two years and ten months.

A large number of observations in the early nineteenth century were collected by James Rennell to form an authoritative and exhaustive work (see Bibliography). The transitory nature of the Gulf Stream was appreciated, and the meandering and the formation of giant eddies with cold patches in their middle were recorded. A distinction was drawn between wind-driven drifts, such as the North Equatorial Current, and streams such as that which jetted out of the Florida Straits and was caused by piling up of wind-driven water. Modern theory makes one system of the whole, but Rennell's ideas led to arguments in 1836 when Arago pointed out that the difference of water level from one side of Florida to the other was only 7½ inches which he maintained was inadequate to produce the fast flow in the Florida Strait. Arago maintained that the driving force of the currents was the difference in density between the Poles and the Equator due to unequal heating by the sun. These ideas were followed by H. Matthew Fontaine Maury, who, although a keen observer and collector of facts was, as Stommel so nicely says, 'very much confused concerning fluid mechanics'.

Maury was superintendent of the U.S. Naval Observatory and may rightfully be considered the founder of American oceanology and one of the pioneers in international studies of the oceans. He collected all wind and current data from ships' log books covering the period 1840–50, and summarized the results in a form which led to the modern system of ocean charts which indicate current flow as well as depth soundings. He organized a maritime conference in 1853 at Brussels in order to co-ordinate observations and rounded off his contributions to the knowledge of the sea by writing his *Physical Geography of the Sea*.

History has a habit of repeating itself, and it is interesting that one of the great experimental and theoretical contributors to our modern knowledge of the Gulf Stream was Professor

Alexander Dallas Bache, who was Benjamin Franklin's great-grandson. The professor was the first man to carry out systematic exploration, and beginning in 1845 he made many drift and temperature observations from the ship *Washington*. He noted hot and cold bands of water, and believed (as some of the most modern oceanographers do) that some of the cold patches were due to mountain ranges on the ocean floor which forced the waters to split into separate streams.

Direct measurements of the surface currents were made by observing the rate of drift of a ship, or by comparing the dead reckoning position with 'fixes' obtained from star sights. Another useful method for following the currents was to cast drift bottles overboard and note where they arrived on shore. The 'Rainbow' expedition in 1802 employed this technique, but the really large-scale use of surface floats was initiated in 1880 by Albert I, Prince of Monaco. Over nine hundred hollow copper spheres were used between the Azores and Newfoundland. The movements of these floats showed indisputably the branching nature of the Gulf Stream, to Norway on the one hand and to France on the other. The first record of deeper current measurements comes from the work of Henry Mitchell, a U.S. Coast Survey engineer in the nineteenth century, who connected two floats with a wire,—one float travelled on the surface, the other sank slowly. Mitchell deduced that the velocity of the current off Fort Chorrera, Cuba, continued to a depth of at least 600 fathoms. Much more detailed observations of the current profile were made in 1885 by Lieutenant John Elliott Pillsbury, who commanded the *Blake*. The captain anchored his ship, and temperature and current were measured across five sections in the Florida Straits.* Pillsbury's excellent measurements are still used as the classical example of how current speeds can be calculated from the density variations, just as wind speeds in the atmosphere are obtained from differences of barometric pressure.

Modern current observers record the salt content of the sea water as well as the temperature, since in most parts of the ocean both these quantities are variable and must be considered when calculating the water density. Direct current

*The ship was, in fact, anchored in many different positions for readings of each of the five sections.

measurements provide confirmation of the movements calculated from the sea-water observations. The end of the nineteenth century saw the development of modern methods of calculation, including a proper understanding of the part played by the forces associated with the earth's rotation. The importance of turbulence in the surface layers of water in determining the transfer of energy from wind to the sea surface was explained following Reynolds' work on flow of fluids in pipes in 1883. In the first quarter of the twentieth century some advance in theory was made by the Swedish oceanographer, V. W. Ekman, but until recently there have been few increases in our knowledge of the Gulf Stream. However, during the past ten years the international approach to oceanography, started by Maury and furthered so admirably by Prince Albert I at the end of the last century, has made it possible to make simultaneous measurements in many parts of the 'ocean river', and to cover the oceans more comprehensively with temperature and salinity observations. Many years ago Lord Dunsany wrote a short story about the confidence trickster who sold the Gulf Stream. It may appear trifling to some that the final bargain was for only ten shillings but we are today at long last beginning to understand the importance of this great 'river', and to assess its true value to the countries whose way of life it affects.

2

THE GEOGRAPHY OF THE GULF STREAM

The waters of the sea are never still. The pull of the moon and the sun cause the tides to wash the coastlines of the world every day, sometimes with a rise of a few feet as in the Mediterranean, but in other parts of the world, where the sea-bed is suitably constructed to encourage the oscillation of the water, the vertical movement is tens of feet, and great stretches of coast are alternately laid bare and invaded by the sea at regular intervals. In northern France, in the gulf of St. Malo, where the tidal rise is in the region of forty feet, these movements of the sea have been harnessed to drive turbines for the production of electricity.*

Wind acting on the sea surface provides the beautiful but awesome waves that may lap idly at the sea-shore, but which also show their fierce nature in storms at sea. The waves of the sea travel at speeds of up to sixty miles an hour, but there is no forward movement of the water. If you watch a piece of flotsam on the sea surface you will discover that it moves up and down in a circular pattern as a wave passes, but it does not get carried forward by the wave—unless, of course, the waves are breaking on the shore, when a general surge forward does take place. The waves on the sea surface are like the waves that can be sent along a rope by jerking one end. A kink travels down the rope, and takes energy to the far end, but the rope is still in the same place when the wave has passed.

But the wind can make water move bodily from one place to another provided it is acting steadily in the same direction for a long time. If you blow across a pan of water a current will be set up. The stream of moving water will be deflected by the back of the pan and will return round the edge. The return current is necessary because the water cannot pile up without causing a head of water which immediately tries to level itself

*A further cause of movements of water is a drastic change in barometric pressure, which can lead to exaggerated tides and wave formations.

by flowing in any direction that is possible—that is, any way that is allowed by the sides of the pan and by the force of the stream of air that is being puffed across the water surface.

The captains of the old windjammers knew that large areas of the ocean were subject to regular winds, and of these the north-east and south-east trade winds have played as important a part as any in the development of the world in the last few centuries. These two steady forces meet in the region of the Equator. This is because the heat of the sun in the Equatorial region warms the air and makes it lighter so that it rises, to be replaced by air moving in from north and south. The rotation of the earth provides a sideways drag to give the resulting winds from the north and south their westerly movement. The north-east and south-east trade winds are the prime movers in the development of the ocean circulation that we call the Gulf Stream. Their steady blow on the surface of the Atlantic causes two separate surface movements, travelling from east to west—the North Equatorial Current and the South Equatorial Current.

If there were no continents, the steady trade winds would maintain their ocean currents circulating regularly around the world.* This happens, in fact, in the southern hemisphere, where the continents of Africa and South America end and where there is a large belt of ocean, stretching right round the earth—it is here that the 'roaring forties' winds develop full force and cause huge waves—waves that are a hazard to lone voyagers such as Robin Knox-Johnston and Sir Francis Chichester, but at least signify a wind which gives a steady push to a sailing vessel. The roaring forties produce a regular circulation between southern latitudes 40 and 60, but they are unique; such a movement is not possible anywhere else in the world, especially in the northern hemisphere, where a glance at the globe will remind one that there are continents, and these cover the greater part of the top half of the world.

The westward-moving Equatorial currents can travel unimpeded for only about three thousand miles before hitting land. What happens when these currents meet the American continent is, like so many natural phenomena, just what one

*Generally speaking, currents are named for the direction in which they flow, winds for the direction from which they come.

would expect. The South Equatorial Current hits the eastern tip of South America at the Brazilian cape—and splits into a southern part which flows down the Argentine coast and a northern stream, the Guiana Current, flowing along the north-east coast of South America, past the mouth of the Amazon, and nearly up to Trinidad. Westward of the lesser Antilles chain of islands this branch meets the North Equatorial Current and together they flow south of Haiti and Cuba almost to the Yucatan peninsula of Mexico. Here the westerly-flowing surface current meets a rebuff, since there is no westerly outlet from the Gulf of Mexico.

The Equatorial currents are forced to move northwards, and skirting around the islands, the piled up water in the Caribbean Sea forces itself out in a gigantic squirt through the fifty-mile wide Florida Straits between Cuba and Miami. At the narrowest part of the strait, the Narrows of Bimini, the flow of the current reaches to the sea-bed at a depth of 1,500 feet and passes through at the rate of ten cubic miles per hour. The speed of the current varies from month to month, but reaches speeds of seven miles an hour in the central and fastest part of the stream. This is a high-speed river, and the total flow through the Narrows has been measured by observing the currents at different depths and at several places across the gap. The mass of water is enormous. At one time, as we have seen, it was supposed that the Gulf Stream was caused by the outflow of the water of the Mississippi river flowing into the Gulf of Mexico. This becomes patently absurd once accurate measurements have been made, since not only is the flow through the Florida Straits an order of magnitude in excess of the flow of the world's greatest river, the Amazon, but on average it carries more than twenty-five times as much water as all the rivers of the world added together.*

This vast jet of water, fifty miles wide and a quarter of a mile deep, aims itself northwards along the coast of Florida. The dimensions of the jet stream are so great that, as we shall see, it maintains its entity for some thousands of miles—like the jet from a hose squirted into a pond it mixes only slightly with the water through which it flows because its volume is so large

*The flow through the Bimini Narrows is about 700 million cubic feet a second.

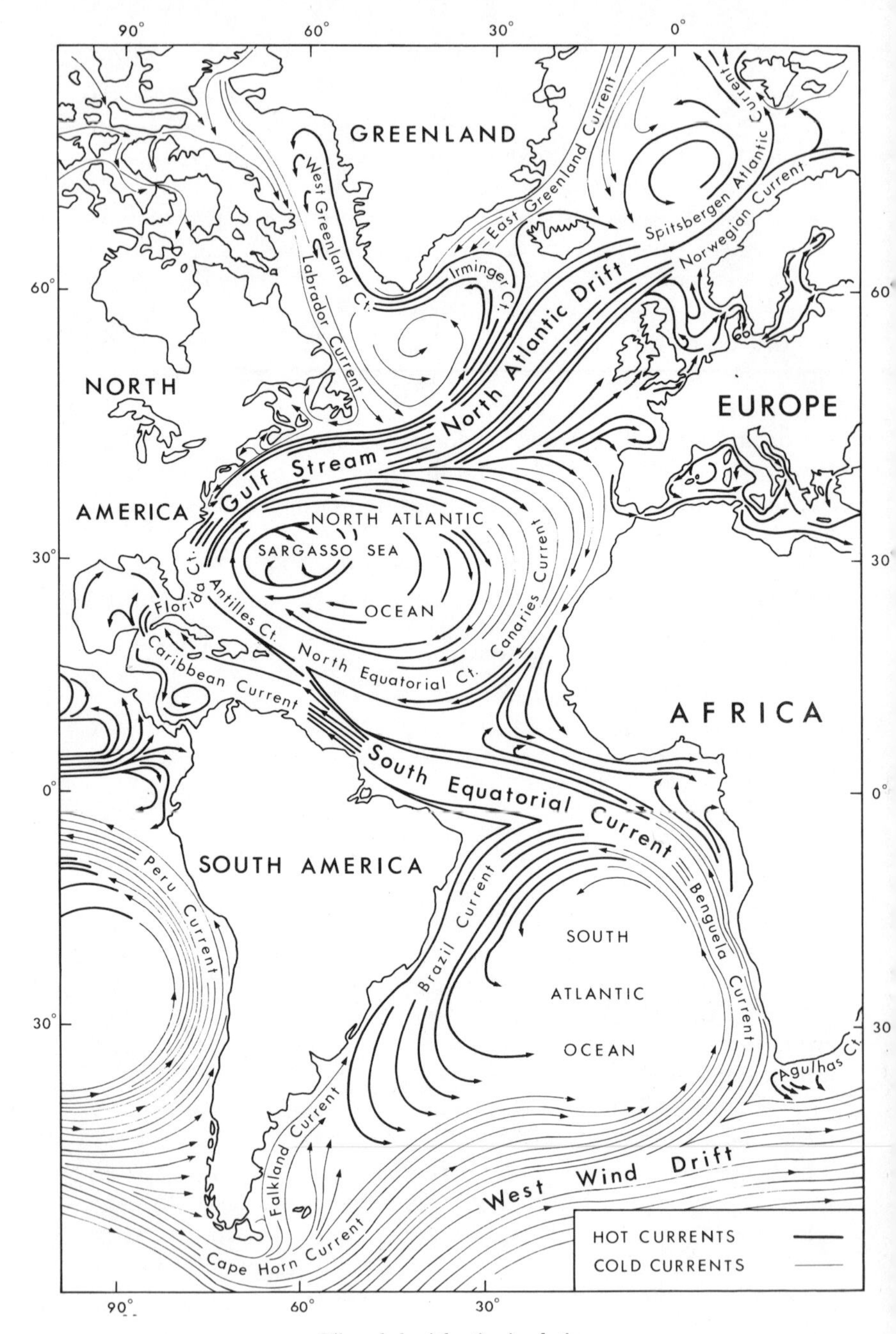

The whole Atlantic circulation

compared with its area of contact. In the same way, the speed gradually gets less, and the fast jet widens out and becomes diffuse on account of eddies and fringe meanderings. However, in phenomena of world size there are other effects which must be considered that are not applicable on the scale of a garden pond.

The fast flow of water through the Florida Straits meets a more gentle circulation which is called the Antilles Current and which is part of a general clockwise rotation of the surface waters of the North Atlantic. The driving force of this circulatory movement is the wind, which in turn is maintained by the rotation of the earth. The earth's rotation helps to give the right-handed motion to the ocean currents. Rotation brings into play Coriolis forces* on anything that moves on the earth's surface, and in the northern hemisphere these tend to give a clockwise movement. The clockwise motion is assisted by winds off the Newfoundland coast region, and probably is augmented by the slope of the sea bed which diverts the stream. In general, however, the rotation of the earth and the winds consequent on this rotation maintain a wheel-like movement of water in the North Atlantic. The narrow ribbon of high velocity water spurting out of the Florida Straits acts as the boundary which prevents this warm water from the Equatorial currents from overflowing the colder, more dense water which borders the North American coast.

As might be expected, the jet stream issuing from the Straits of Bimini merges with the main wheel-like circulation of the North Atlantic waters. The hub of the wheel is the notorious Sargasso Sea, where ships were supposed to be captured in masses of seaweed and where a solid land-like region of hard-packed seaweed was supposed to exist. You can, in fact, sail through the Sargasso Sea without impediment, and without seeing much more evidence of the stagnant centre of the whirling Atlantic than some trails of yellow seaweed. This does not mean that the Sargasso Sea has no importance. It may well be the breeding ground of the world's eels and it certainly has a geographical significance in being the axis of the Gulf Stream circulation.

*Coriolis forces are a natural inescapable phenomenon caused by the rotation of the earth, and applicable and noticeable in any large body of fluid, air or water.

We are not yet sure what makes the Gulf Stream kick eastwards across the Atlantic. There may be a push off the Newfoundland coast, provided by the south-flowing cold Labrador Current. Perhaps the Coriolis forces play a part, helped by the sea-bed rises and ridges which act as deflection plates; the movement of the Gulf Stream is much more easily explained in terms of mathematical formulae than in words, but these formulae change with the years and common sense provides the main overall picture. Whatever the reason, the drift of water swings eastwards across the Atlantic, to provide Benjamin Franklin's Gulf Stream—his well advertised navigational hazard. The force of the current diminishes as we go east across the Atlantic and on reaching Europe the more diffuse stream splits. Part must follow the wheel-like circulation of the whole North Atlantic body of water, while a segment reaches up to trail around Scotland, finally to die out around the approaches to Spitsbergen. On the way this latter stream of warmish water gives solace to the north-west Scottish coast, providing palm trees at Ullapool and pleasant warmth at Oban.

In a way Maury was wrong in calling the Gulf Stream a river, and oceanographers try to forget this naïve explanation. However, there is no doubt that the combination of the hosepipe jet effect at Florida with the general circulation of all the North Atlantic ocean waters does provide a body of water which keeps its entity and which flows river-like around the ocean. 'There is a river in the ocean', as Maury starts his book (see Bibliography), is not a bad description of what happens. Professor Jacques Piccard managed to drift in this particular river for a month in his epic voyage of 1969 (see Chapter 6). The Gulf Stream behaves like a river in one important respect, and this made life difficult for Professor Piccard's journey.

Just as a river flows slowly across a plain, after leaving the rapids and gorges and waterfalls of its high-speed young life in the mountains, so the Gulf Stream in its more mature phase loses its clearly defined boundaries and pursues a more diffuse life. The Gulf Stream meanders, as a river will do when it reaches a flat plain; obstacles direct its path easily to right and left, and in the course of time great loops take the place of a straight arrow-like direction. Sometimes the loops of a river, which accentuate themselves because the erosion of the river bank is greatest at the outside of a bend, become cut off in a flood period to leave ox-bow lakes. This lazy phase of a river can be seen clearly in many parts of the world now that air travel has made it possible, when cloud cover disappears, to take a bird's-eye view of geography.

The Gulf Stream starts its gentle life a few hundred miles north of Florida, and careful tracking of its path has shown a large meandering of the main river, together with eddies which swirl off to one side and the other and which sometimes end up as local, temporary circulations of water. The jet stream is losing its energy, at the same time mixing of the body of water which spurted out from the Florida Straits is taking place, and gradually the high-speed current becomes a more gentle outer rim of the main North Atlantic wheel-like circulation. This does not mean that the Stream is dead. The one- or two-knot current which worried Franklin so much because it slowed down the mail packets, continues in a west-to-east direction across the North Atlantic.

3

THE MOVING SEA

The Gulf Stream is not the only current in the oceans, even if it has gained a notable position in the western civilization which it has served so well. The whole three-mile-deep stretch of water, covering two-thirds of the globe, is in a constant state of turmoil which is just as well for mankind. Movement is to be expected, since water finds its own level, and if the winds caused by the earth's rotation move the surface waters, some return flow must take place to avoid piling up of material downwind. The return currents may circle around the sea surface, but they also travel deep, sometimes along the floor of the ocean, three miles below the surface. There is, in fact, a 'grand circulation' of water masses in the depths of the oceans—not nearly such a rapid movement as that of the Gulf Stream, but nevertheless a steady and regular system of transport of water from one part of the ocean to another. At first sight one might expect the warm Equatorial waters to rise and the cold Polar waters to sink, to give a simple circulation symmetrical about the Equator, with cold water travelling along the ocean bed from the Poles and warm water replacing it by flowing along the surface. This is the pattern we would get if we heated a pan of cold water. However, the ocean heating is at the surface, not at the bottom of the pan, so that this simple circulation picture occupies only the top few hundred feet of the sea, and is confined to the central part of the globe. A secondary effect on the Equatorial and sub-tropical circulation is the effect of increased salinity due to water evaporation. The salty water is heavier and tends to sink. Overriding those fundamental circulations of temperature and salinity are the wind forces which, as we have seen, account for the Equatorial currents and are the driving mechanism of parts of the Gulf Stream.

It is easy to see why some deep currents arise. Ice melting in the Arctic and Antarctic regions forms cold water which, being heavier than the warmer surface layers, slides down gently away from the Poles. The Polar deep currents become spread out as they travel, simply because the lines of longitude, which meet at the Poles, diverge all the way to the Equator. The deep Polar currents thus tend to lose their entity for purely geographical reasons. They never in fact reach the Equator, where presumably the north and south currents would in some way nullify each other, because of opposition from other currents encountered on the way.

The main deep currents then, consist of layers of colder water which sometimes appear at the surface, in particular in the Polar regions, and exist at depth in all latitudes. The boundary between this cold deep water and the warmer surface circulation is not well defined, but generally oceanographers consider 10°C. (50°F.) as the temperature which is the dividing line. This boundary is shallowest near the Equator and becomes deepest around 30°N and 30°S, beyond which it moves up to reach the surface in Arctic and Antarctic regions. The deepest layer of warm upper water approaches 3,000 feet (915 metres) in the area of the Sargasso Sea, which is the middle of the main North Atlantic circulation, and is a quiet but interesting part of the Gulf Stream mechanism. Chapter 7 describes how the upper layer currents circulate around the three great oceans, but first what is happening in the colder water down below the familiar surface currents?

Before the first transatlantic cable was laid few people thought much about what lay beneath the sea surface, and many of those (except for the minority who believed the oceans to be bottomless—how, one cannot imagine) conjured up pictures of gigantic sea serpents and other formidable creatures inhabiting a dark, silent, murky abyss, where the familiar surface motions of waves, tides and currents were completely stilled. These illusions were dispelled when cables were successfully laid, and in some cases recovered with small animal life attached, and when samples from the sea-bed and deep current measurements and all the wealth of modern oceanographic information were gathered. The picture of a massive deep-water circulation gradually became apparent to physical oceanographers of the 1920s and the mechanism of this circulation is now roughly understood. The circulation of the deep waters is a slow process compared to the water movement apparent at the surface in rivers, or in currents such as the Gulf

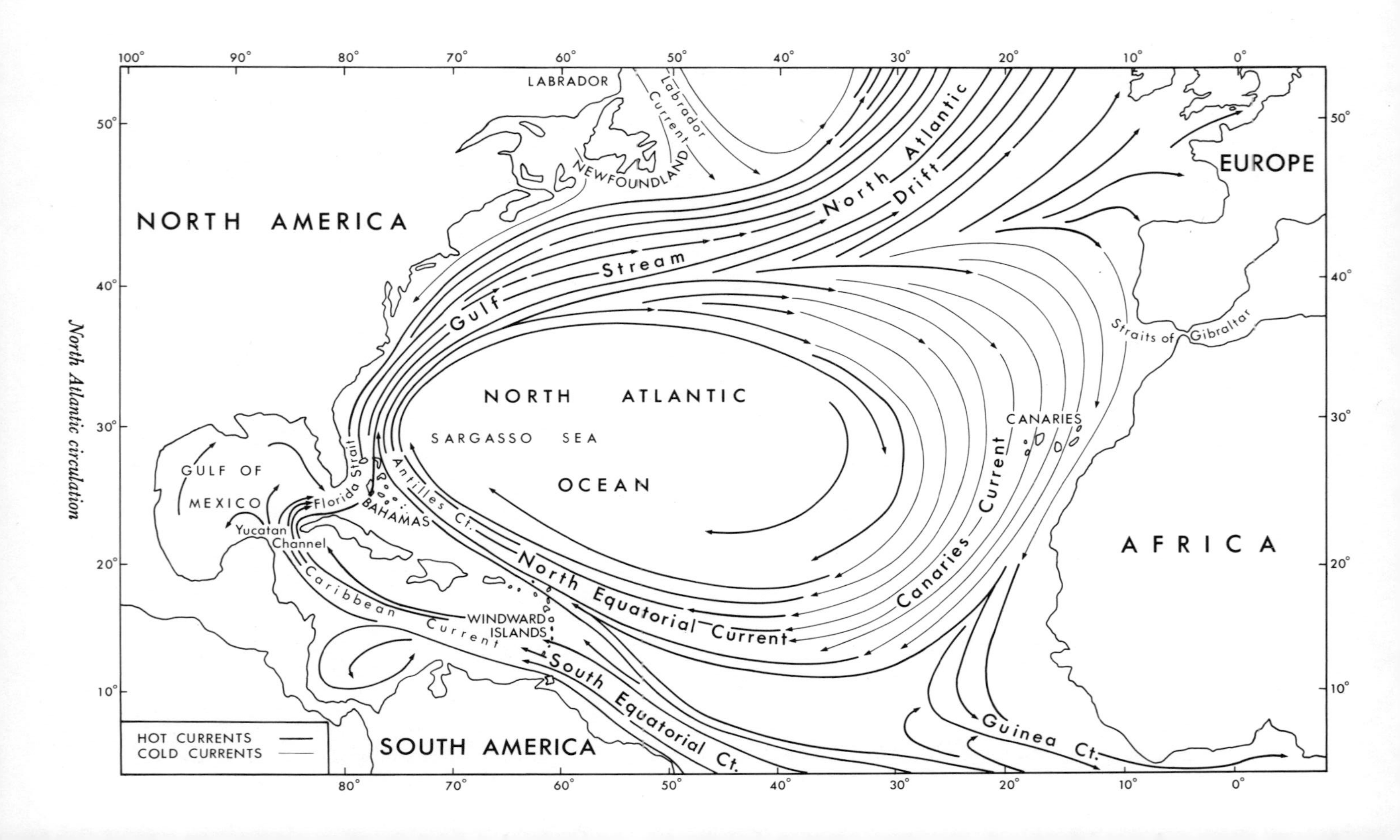

North Atlantic circulation

Stream, but it is rapid compared with another time scale that we think of these days. The geological processes that have shaped the earth occur in millions of years. The deep ocean currents make their circuits in hundreds of years, and at those rates can have a noticeable impact on the general ecology of the seas, just as the slow but inexorable flow of ice down a hillside provides the gouging of the earth's face by glaciers.

The prime mover in the deep-water circulation is the sinking of cold Polar water that has been mentioned several times in discussing the warm surface currents, and the controlling factor for the detail of the deep currents is the shape of the sea-bed, in much the same way that the shape of the continental basins, and the land barriers determine the flow at the surface. The Arctic bottom water is almost locked in by land, since during the past 100 million years or so the continents have drifted to their position which is preponderantly in the northern hemisphere. Some cold, heavy deep water slides out between Greenland and Spitsbergen, but the Scotland–Iceland–Greenland ridge does its best to limit this southerly flow. The bulk of the deep ocean water, both in the northern as well as in the southern hemisphere, stems from the Antarctic. Here we have almost a mirror image of the northern hemisphere. A large southern ocean surrounds a Polar land mass, opposing the continents of Europe, Asia and North America which form the periphery of the Arctic Sea. Antarctica provides a storehouse of ice nearly two miles in thickness and a few million square miles in area, collected slowly, but in historical times, by the heavy snowfalls that take place in the long winter months. This enormous reserve of cold forms the refrigerator for the surrounding sea-water; by means of icebergs broken off the tongues of glaciers the sea is cooled. The cold water is heavy compared with warmer surface water, and a little boost to its density is provided by the increase in salinity which occurs when pure ice crystallizes out at the surface.

The heavy cold water production is at its most efficient working in the Weddell Sea, and in this marginal area of the Antarctic continent, the Antarctic 'bottom water' originates. This bottom water glides slowly downwards along the continental slope and spreads northwards over the floor of the Atlantic, Antarctic and Indian Oceans. The subsequent history differs from one ocean to the other for similar topographical reasons that (see chapter 7) provide different surface current patterns in the three main oceans.

The cold water which sank to the bottom in the Weddell Sea flows steadily northwards past the Equator to about 40°N in the Atlantic Ocean. However, the flow is unsymmetrical in that there is a stop on the east side caused by the shallow Walvis Ridge which stretches out from the south-west coast of Africa. The fact that this ridge acts as a barrier can be deduced from the temperature measurements of the bottom water; to the north of the ridge it is distinctly warmer than to the south. The bottom water to the north of Walvis Ridge arrives by a most circuitous route; it is Antarctic water which has travelled northwards up the west side of the Atlantic and has cut through the barrier of the mid-Atlantic ridge at the gap provided by the Romanche Trench. This arm of the main flow turns southwards, having warmed a little on its journey, to provide the deep water north of the Walvis Ridge. A little Arctic cold bottom water escapes over the sill either side of Iceland to make a small contribution, but the greater part of the Atlantic bottom water comes from the Weddell Sea in the Antarctic.

In the Indian Ocean the bottom water is supplied from the Antarctic, but some also flows in from the Atlantic. The Red Sea behaves in a similar way to the Mediterranean in providing some salty deep water, but the flow is only about one sixth of that through the Straits of Gibraltar. There is, of course, no possibility of Arctic water in the Indian Ocean because it is closed by land in the northern hemisphere.

The Pacific Ocean is almost shut off from the Arctic Ocean by the American and Asian continents, the gap at the Bering Straits being narrow and shallow. The Weddell Sea is once again the supplier of bottom water. There are no salty water contributions as from the Mediterranean or the Red Sea, and no cold sources as in the Atlantic, so that the deep layer in the Pacific is not a very distinct phenomenon. Some deep water flows from the Atlantic, but it becomes mixed with the bottom waters. An intermediate source of water exists in the Pacific between the deep water and the warm surface waters. The flow is northwards from Antarctica in the southern hemisphere and south from north-east of Japan in the northern hemisphere.

The quiet deep ocean currents are most important to the world today. The circulation of the sea stirs up the waste that is dumped in the ocean and allows it to become so dilute that it is virtually lost. Now that we are making radioactive wastes we wish to know whether the mixing will occur adequately to avoid any pockets of dangerous concentration. The overall circulation rate between bottom and upper water layers controls the rate of mixing to a large extent, and this is a figure that we must know if we are to do our waste-dumping safely and sensibly. Early estimates of the rate of flow of the deep water layers were obtained from calculations based on the pressures measured at different parts of the ocean; from these pressures a speed of current can be calculated in a similar way to estimating wind speeds from differences in barometric pressure. It appears from these calculations that the transport of the northern deep water of the Atlantic takes about five hundred years to complete its journey, while in the Pacific the time is longer because its deep circulation is not very clear-cut in relation to other bodies of water, being an offshoot of the Atlantic deep circulation.

We can take samples of the deep water and measure their age by modern radioactive techniques such as that based on Carbon-14. The decay of this particular active form of carbon is known, and it is also known that there is a steady supply of radioactive carbon to the atmosphere, and hence to the surface waters of the sea. Once the Carbon-14 is trapped in the sea surface waters it decays; in about 6,000 years only half the active atoms will be left, and so, by measuring the Carbon-14 concentration it is possible to work out how long the sample has been in the sea. For the mid-Atlantic Ocean the deep water has a Carbon-14 age estimate of 650 years, while the Antarctic bottom water is 900 years old. Greater ages are found in the Pacific, ranging from 1,300 to 2,000 years, agreeing with our knowledge of the circulation of deep waters, with the Atlantic the main source. It is interesting that there is a significant increase in age from south to north, the figure being 600 years over a distance of 4,500 miles, corresponding to a movement of about seven miles a year—not fast compared with the way people move around, but a rapid flow compared with geological changes.

Other deep currents in the ocean are caused by the sinking of bodies of heavy *warm* water. This is because evaporation in hot climates leaves the surface sea water more salty, and the presence of above normal salt dissolved in the sea makes the water abnormally dense. Then again, there are deep currents which are merely the return flow corresponding to a surface circulation, such as the North Equatorial Drift which, we have seen, is a section of the North Atlantic Gulf Stream pattern. Once set in motion the deep currents move inexorably on for thousands of miles, since they consist of a great volume of water, which only diffuses itself into the rest of the ocean by spreading sideways and by mixing at its top and sides. If the current is of large dimensions the zones of mixing are small compared to the body of the current. Thus we find, both with deep circulation and with currents at the surface, bodies of lighter or denser water maintaining their identity for very great distances. Since the qualities which determine whether a water mass is light or heavy are the temperature and salinity, it is possible to label particular bodies of water by quite simple measurements. It is for this reason that oceanographers take the temperature of sea water, not only at the surface but at regular intervals to the bottom, and also collect samples of water from different depths so that the salt content can be determined in the laboratory.

An example of a heavy salt current is provided by the Straits of Gibraltar exit from the Mediterranean. The Mediterranean is a warm sea, with only one outlet to the Atlantic Ocean and the surface evaporation tends to produce water with excess salt. This water, being heavier, takes the bottom position at the sill of the Gibraltar exit, being overlain by lighter fresher Atlantic water. There is a build-up of water in the Mediterranean due to the supply from the rivers which empty into the sea, so that a net flow outwards of water must occur at the Straits of Gibraltar in order to maintain levels correctly with the Atlantic. The outflow, because of the heavy salty water predominating in the Mediterranean, takes place in the bottom section of the profile of the Straits, and there is actually a flow inwards at the surface through the Straits. This surface inflow from the Atlantic no doubt gave birth to many of the old myths about Atlantis, by washing unusual flotsam in from the Atlantic, and some delvers into old Greek and Roman writings have associated the currents around Gibraltar with the difficulties of crossing the Styx.

Navigation of deep submarines is similar to navigation at the surface of the sea, and it is affected by the ocean currents. Nature plays some odd tricks which have been used in navigation during wartime. The deep current of heavy salty water that slides over the sill of the Mediterranean at the Straits of Gibraltar was used by Italian submarines as a conveyor belt to carry them silently past the acutely tuned listening devices on either side of the Straits. An intermediate layer of water is recognizable as a deep flow between the warm surface water and the cold bottom water. This behaves in the reverse way to the bottom water, in that its source is in the Arctic and it flows far down into the southern hemisphere. The high oxygen content of this deep water indicates that it started at the surface, and on tracing the current back to its source, it is found that the Gulf Stream is what one might call one of the founder members. Where the salty Gulf Stream mixes with the cold Polar water off Greenland, there is strong cooling in the Irminger Sea between Greenland and Iceland, and in the sea between Labrador and Greenland. Winter winds help to cool the sea in these areas and produce a layer of heavy water which sinks and flows southwards. A third source of this deep saline water is picked up off the Straits of Gibraltar where the heavy salty water slides out beneath the warmer surface water. When the Antarctic continent is approached the intermediate water is pushed to the surface by the even heavier bottom water flowing in the opposite direction from the Weddell Sea. The deep water brings with it plenty of nutrient from the deep ocean and this causes a prolific growth of plant and animal life in the rough Antarctic waters. This is why whales, who need an enormous supply of food, congregate in the southern latitudes in the feeding season.

When a current moves up to the surface after travelling along the sea-bed, it acts as a conveyor belt to bring rich nutrients to the near-surface layers of water. These nutrients consist of the decayed remains of sea animals and plants which inexorably fall to the bottom of the sea when the plants and animals die. They are as necessary to further life in the sea as are fertilizers and trace elements in farming on land. Even before the days of artificial fertilizer—which consists mainly of nitrogen, potassium and phosphorous compounds—farmers fed the land with animal manure and rotated their crops to ensure a balance between what one crop took out and its counterpart replaced.

The rule of life in the sea is that big fish eat little fish, and so on down the line, which ends at minute organisms which float freely in the top few hundred feet of the sea, and which thrive and multiply on a diet of carbon dioxide and sea-water. These beginnings of the food chain organisms need light and they breed faster when the water is warm. It has always been a paradox, therefore, that there is often very little fish life in the warm, bright Equatorial waters, and that it is the Polar regions where the vast tonnages of fish are caught, and where whales go to feed on the thick soup of shrimp-like animals known as krill which forms their staple diet. The answer is, of course, that the warm Equatorial waters form a stable shallow surface layer, because the warm water is lighter than normal. If any fish or plants did ever live in this layer, their remains at the end of their life would sink through the underlying waters to the sea-bed, and no 'fertilizer' would remain in the warm, bright surface layer. It is only when currents move from the sea-bed, as they do in the Polar regions, and bring the necessary additives to the surface layers, that plenty of life can exist.

In places like the North Sea, where the famous Dogger Bank has long been a favourite fishing ground, the shallowness of the sea allows stirring up by waves and tidal currents so that there is plenty of nutritious bottom material to add to the carbon dioxide and light necessary to maintain growth. There are some places such as the sea off the coast of West Africa by Walvis Bay, where bottom currents appear at the surface, having ridden up on top of an opposing water movement to form rich areas of 'upwelling'. With the abundant light and plenty of bottom nutrients, Nature sometimes overplays her hand and disaster occurs. Too much growth due to the excellent conditions for the start of the food chain may cause too many plants and animals to be produced, to such an extent that all the available oxygen dissolved in the sea-water is used up. Fish then die, and get washed up on the beach in droves. The decaying fish at sea putrify and form hydrogen sulphide which makes the anaerobic conditions—i.e. lack of oxygen—even worse. Similar cycles of nature take place in some rivers where life is stimulated by nutritious effluent from sewage systems and by warm water from power stations. The result of anaerobic conditions in a river are unpleasant and is one of the world's worries now that population has increased so rapidly. However, like most things in the world, man is only copying Nature, and by studying how the natural balance is restored in the oceans when over-production occurs, it will be possible for scientists to show how to control the bad effects on a smaller scale that man produces.

The sea has always been one of the traditional dumping grounds for man's waste. The volume of sea-water on the earth's surface is about 300 million cubic miles, so that there is adequate volume to absorb a vast amount of waste.* There is one proviso, however, and that is concerned with mixing. If the diluting effect of the whole ocean is to be used there must be adequate distribution of the waste throughout the sea. The stirring of the seas is carried out by the ocean currents, and this is one of the reasons why today especially, with big pollution problems arising, it is important that the effect of currents should be thoroughly understood.

One can see what happens when too much of man's waste, whether from industry which supplies his needs, or from his own domestic effluent, goes into too little water by looking at the sad case of Lake Erie. Lake Erie is a large body of water and when settlements first grew around its shores its absorbing capacity for local sewage was quite sufficient. The bacterial action and the animal life in the water readily used up all the waste that was piped into the lake. Even in the early days of industry, sulphite wastes from paper manufacture disappeared into the lake with no obvious harmful effects. It is probable that they too were degraded into a final useful nutrient for fish life. As the population and industry gradually increased, the ability of the lake to cope with the increase of material gradually waned. Warning signs were noticed by biologists, but citizens did not want an increase in rates to pay for new sewage plants, and many old industries had neglected to put aside any reserves to face the day when they would have to treat their effluents

*One cubic mile of water weighs about 5,000 million tons.

before returning the water to the lake. They would either have to carry on in the old way, or go out of business.

The position has worsened, not only with population explosion and its increasing demands for material goods which must be manufactured, but also because of a subsidiary consequence of these two factors. More mouths to feed has meant more intensive use of agricultural land by the use of fertilizers, while the easier life has encouraged the use of domestic liquid detergents in place of solid soap. The run-off from the land and the domestic waste-pipe have fed the remnants or end-products of these materials into the lake and both have helped the plant growth. This might at first sight be an admirable thing if only the rather delicately balanced natural cycle had not been overworked. The increase of growth has occurred all right, but at the 'beginning' of the food cycle, with algae. These green weed-like plants provide food for fish, but with too much growth they take all the available oxygen in the water, and so deprive the fish of something which is just as essential as food.

Lake Erie is now in an anaerobic (or oxygen-lacking) condition, as are many rivers in many parts of the world. Sometimes the excessive waste-dumping has been assisted in causing pollution by the pouring of warm water into rivers, the water having been used to cool large power plants. The warm water is a growth promoter, and although experimental schemes are being tested to carry out fish farming in the warm-water effluents of nuclear plants, uncontrolled 'dumping' of warm water in rivers can lead to anaerobic conditions.

Lake Erie and the polluted rivers can be made clean again, but this clean water must be paid for. The bill for modern sewage plants will fall on the tax-payer, and the cleaning up of industrial effluent will cause an increase in produce prices. There are probably many people who would prefer dirty water to a decrease in material comforts, so that improvements may well be slow since money must be extracted from a public which is sometimes reluctant to pay for them. However, the trouble began with human beings reproducing too quickly and wanting all the good things that modern industry has made available. Man must learn to advance a little more slowly if he is not going to live in ever dirtier conditions.

The fossil fuels—coal and oil—cause pollution of the air

with black smoke and automobile exhaust fumes. The problems are being tackled by phasing out the use of coal and by cleaning exhausts with such devices as afterburners. There is a demand for greater use of electricity derived from nuclear power but this in turn may introduce new pollution worries. When the first British reactor was built in Cumberland in 1956, it was necessary to put dilute radioactive material into the sea, rather than pump it into the ground and risk contaminating the water supply. The need for achieving sufficient mixing before there was any chance of the material returning to shore was realized and a comprehensive set of current measurements was made in the Irish Sea to determine the best point offshore to inject the waste so that it would be carried out to mix thoroughly rather than be returned to the shore.

Another potential danger of radioactive waste has appeared. Some fairly 'hot' material has been sealed in containers and dumped in deep water. Although theoretically the containers cannot leak there is always a chance of human error or even of some remote accident such as an explosion going off near the containers. Careful current measurements are therefore essential to ensure that in the unlikely case of leakage the waste would not be taken anywhere where it could be a danger. For this type of planning the whole deep-water movement must be known in addition to the surface currents and those in shallow offshore waters.

Another accidental type of spillage in the sea is provided by the example in 1967 of the *Torrey Canyon*, an oil tanker which ran aground off south-west England while carrying a cargo of crude oil. The oil drifted with wind and surface currents on to the shores of Cornwall and Brittany, and caused consternation amongst the local populace, especially since the sinking took place a few months before the summer holiday season, on which the livelihood of many of the local inhabitants is based. A co-operative effort involving government research departments, the Services and the oil companies was successful in cleaning up the beaches. In the British case large quantities of detergents provided the cleanest beaches holiday-makers had seen for years. Unfortunately the detergents used were lethal to all small animal and plant life living between high and low water. All the limpets, mussels, seaweed, and so on were removed with the oil and other dirt. Although this caused an outcry by many marine biologists, things were not as bad as appeared at first, since all the marine animals were re-established within two years. The action taken was the right one since these were beaches used by many human beings. There has now been an improvement, since the new oil dispersents are much less toxic and new precautions against accidents have been established in the oil industry. However, we must remember that man does have a rightful place in the world, just as much as do other animals. There is a tendency today to denigrate man compared to other animals; in fact he is much better than most, not nearly so cruel as wolves, for example, or so mean and predatory as many insects. Other animals also spoil their habitat; there is a wood in Mount McKinley Park in Alaska where all the trees are dead because a wasteful group of porcupines scraped off the bark. The wood will take years to recover, and if man had behaved as the porcupines did he would have been severely condemned by many conservationists. We are much more likely to produce a better cleaner world if industry and ecologists work together sensibly and if extremists stop criticizing wildly without first ascertaining facts about the magnitude of any industrial impact on the environment.

Ocean currents are, then, important these days both for harvesting fish from the sea and for using the sea as a controlled and where possible productive rubbish-dump. There is a third important reason why the pattern of ocean circulation should be thoroughly understood, and this has a longer-term significance than dealing with world hunger and waste disposal, the two modern challenges arising from man's sudden population take-off. The more we learn about the weather and how changes are triggered in respect of such phenomena as hurricanes, the more we find that the exchange of energy between wind and water is necessary to a full understanding both of local weather conditions and of long-term climatic changes. The currents, such as the Gulf Stream, are the great stirrers of the oceans, and a knowledge of their influence is all-important in the forward planning of man's activities if he is to think at all to the future. It is a poor reflection on our morality that we prefer to spend money and effort on things that will increase our material well-being, or on satisfying our curiosity and providing exciting spectacles by achieving firsts in the way of exploration, rather than by collecting information about how the forces of Nature work. Current measuring at sea is not terribly exciting work, and it requires expensive ships and a great deal of effort. It is difficult to persuade people and governments that a large effort is worthwhile when the consequence of not making the measurements now will fall on future generations rather than on ourselves. However, we shall not be thanked by our successors on this earth if we have left them a series of problems that call for drastic measures, but have not taken the trouble to understand the basic mechanisms of climate changes or fish growth or dispersal of waste. Engineers can always cope with problems, however large, provided they have the basic data. We should divert far more of our effort now to collecting data about natural processes and spend less effort in construction of new things. This may mean a temporary standstill, a sort of consolidation period in our development, but it will be appreciated by our children and grandchildren.

4

CLIMATE AND METEOROLOGY

The Gulf Stream is driven by the force of the wind on the surface of the water. The earth is a watery planet, the most beautiful we know so far, as the photographs taken from the moon have demonstrated. The blue and white appearance of the earth to space travellers underlines the two influences that affect the climate and weather enjoyed by the inhabitants of the earth. Blue for water and white for cloud—the sea, occupying two-thirds of the earth's surface, and the cloud pattern, sometimes providing welcome shade, at others, the source of wild disturbances that can cause catastrophe to the minute beings that crawl on the ground.

Air and water on the earth interact in other ways than the direct effect of steady winds turning the surface layers of water to start a regular ocean current. The sun beating down on the sea surface warms the water, and in turn a warm air layer is produced. The warm air is lighter than the colder air above it, and tends to rise. A circulation pattern of the air is started. The sea-breeze at the seaside is a familiar example of how air circulation is produced by heat from the sun being absorbed at the earth's surface. The air itself is too thin to be a great absorber of energy, and therefore has to rely on land and water to collect its heat for it, and transfer it by the interaction of air with the warm surface. Land and water behave differently in this respect: land is solid and fixed, and is a very poor conductor of heat, so that it warms up quickly when it absorbs the sun's rays. On the other hand, the surface layers of the sea are constantly stirred so that the heat absorbed by the surface is shared with water underneath, and, other things being equal, the sea warms up less quickly than the land. This accounts for the gentle sea-breeze that develops on a summer's day. The land heats up rapidly, and the hot air above it rises, to be replaced by cooler air flowing in from the sea. The movement of the air stream across the sea surface may produce a little extra cooling by evaporating water during its progress to the shore. The whole mechanism makes life for mankind much more tolerable, and it is certain that human ingenuity could not have devised a cheaper or more efficient system.

There are times when the exchange of energy between water and air can lead to gigantic phenomena involving amounts of energy as great as that contained in a thousand-megaton bomb. Hurricanes start over the oceans in quite a mild way with a gentle circulation of the air. However, the moving air, with its base associated with the sea surface, soon assumes the aspect of a giant engine. Sea-water is evaporated from over a large area and is swept upwards. The water condenses and in so doing provides a new source of heat; the air is warmed further and accelerates its spiralling rise. Soon the wind strength increases to 100 knots and there are violent rainfalls. The more the engine speeds up its rotation, the further afield it draws air laden with moisture from the sea surface. The only quiet place to be in a hurricane, and this is only a temporary respite, is in the centre of the swirling masses of air. Here, in the 'eye' of the storm, usually an area of about fifteen miles in diameter, the barometric pressure is low, winds are light and variable, the air is descending and warm and there may be a little cloud. When the barometer starts to rise again, the full force of the winds begins once more, and the second half of the storm must be ridden out.

Hurricanes are a hazard at sea, and often do immense damage on land before they dissipate the energy that has been collected by the interaction of air and warm surface water. Active research has been carried out for many years in an attempt to switch off the air-sea engine before it builds up too much power (for example, by causing rain at an early stage) but no certain success has yet been achieved. However, the weather satellites, which for the past ten years have been photographing the pattern of cloud covering the earth, are not only showing how disturbances such as hurricanes begin, but are also teaching more about the general air circulation around the earth.

The other side of the weather picture is concerned with the stabilizing effect that the sea exerts on the heat balance of the earth. The currents, such as the Gulf Stream, are the great ocean stirrers, and a knowledge of their influence is all-important in the planning of man's activities if he is to think at all to the future. For many years meteorology, especially that aspect of it concerned with predicting the weather, had to rely on statistical methods of forecasting, based on observations taken in the past, and on deductions by comparison with similar geographical conditions.

A great advance in elucidating the mechanisms of depressions, fronts and other phenomena that determine our weather is today becoming possible, now that it is practicable to see what is happening by viewing from the outside, instead of looking upward at the underside only of a bank of clouds. It is increasingly important, therefore, that the ocean circulation should be thoroughly understood because the speed at which the sea can absorb excess energy from the air and the extent to which the whole volume of the sea acts as a reservoir of heat are among the unknowns in the mathematical and physical models of the atmosphere that are being constructed in order to work out the problem. The circulation of the ocean waters is affected by the earth's rotation, and by the steady winds at the earth's surface. These winds in turn depend upon the rotation of the earth. Just as the ocean currents tend to circulate clockwise in the northern hemisphere and anti-clockwise in the southern, so do the cyclonic disturbances in the atmosphere. This is no mere fortuitous happening, it is a consequence in both cases of the direction in which the earth revolves on its axis. As we learn more, we shall regard both air and sea as part of a world circulation and a heat balance picture. In the past, the methods of deducing where currents flowed were dependent on similar measurements to those of barometric pressure, which determine wind direction in the atmosphere. While these measurements of ocean circulation continue, new observations to measure the detail of exchange of material and heat at the sea surface are being made so that we may progress further with our proper understanding of these natural phenomena.

From a more mundane point of view, forgetting the fundamental scientific aspects for a while, the Gulf Stream has always been notable for its almost direct effects on climate in various parts of the earth. Unfortunately for North Americans, the warmer, faster-flowing part of the Gulf Stream keeps its distance from the shore, and, in fact, there is a south-moving counter-current running along the eastern seaboard of the

U.S.A. The existence of a stream of warmer water a few hundred miles offshore can in many cases lead to an amelioration of climate, but in the case of the U.S.A. there is another sad fact to stop any warming effect of the Gulf Stream. The American continent is on the wrong side of the Gulf Stream, and there is nothing, apart possibly from reversing the direction of the earth's rotation, that can be done about it. The prevailing winds are from the land to the sea in North America, so the warm Atlantic waters cannot provide the heat to form benign winds to improve the winter climate of the north-eastern states of the U.S.A. This is why the winter in the New England states is much more severe than that of the south of France or northern Spain, although they are all at the same latitude.

There is no doubt that, for warm winters, it pays to live on the west side of a continent. North-west Europe is bathed by the remnants of the northern branch of the Gulf Stream, and so gains a few degrees in water temperature. This keeps the ice from moving far south in winter, so that Norway is virtually ice-free, even in latitudes of 70°N, while the Baltic Sea, far to the south, freezes over in large areas every year. More important, however, than the direct surface-water temperatures, are the warm winds from the south-west, which are generally the prevailing winds, but whose movement is assisted by the low pressure atmospheric conditions created by the warm water to the west of Norway.

Winds blowing over the sea surface, which, provided the sea is not frozen, will always be at a temperature above 0°C. (32°F.) are always warm winter winds for a similar, if opposite, reason why the sea-breezes are cool in summer. The sea is stirred, and acts as a heat reservoir and a stabilizer of temperature; the land surface heats up quickly, but just as rapidly descends to below freezing point when the sun's warmth is withdrawn. In the case of north-west Europe, the exceptionally warm sea surface provided across the Atlantic augments this warm air circulation, but it is the air circulation rather than the warm water which provides the mild weather at places like Ullapool on the west coast of Scotland, and which allows palm trees to flourish in those northern latitudes. If the Gulf Stream circulation were to stop, those palm trees would not necessarily perish, provided the south-west warm winter air stream continued to blow from the temperate latitudes of the Atlantic Ocean. As we shall see later, large climatic changes which have occurred in the past have beginnings and endings which are not yet fully understood, and which may have many small causes.

There is another well-known climatic effect of the Gulf Stream. This occurs before the Stream moves over to perform its pleasant task of warming Britain and Scandinavia. Near Newfoundland, where the circulating current is swinging to the right before traversing the North Atlantic Ocean, the stream of warm water collides head-on with the cold Labrador Current. To what extent the Labrador Current pushes the Gulf Stream around to an easterly direction is not certain. Modern research is still endeavouring to assess the relative importance of the several mechanisms which can deflect this ocean current. There could be a contribution from the change in topography of the sea-bed. The Newfoundland Bank could form a barrier to the deeper part of the current flow and thus help to deviate the main stream. On the other hand, it is fairly certain that the Coriolis forces, arising inexorably from the rotation of the earth, would give a clockwise circulation in the northern hemisphere of any ocean, regardless of other forces.

The Labrador Current is the left-hand member of a pair of outlets from the Arctic Ocean. The Gulf Stream, whose history is very much bound up with the ice and water movements in the Arctic Ocean, pushes water along the coasts of Norway and Spitsbergen. This water cannot pile up indefinitely, any more than it can in the Gulf of Mexico as we saw in Chapter 2, and the Labrador Current to the west of Greenland, and the Greenland Current along Greenland's east coast provide the outlets. The Greenland Current sticks close to the shore, and is joined before the southern tip of Greenland is reached, by a finger from the Gulf Stream which at this point has pushed northwards to bathe Greenland in comparatively warm water. The small tributary of the Gulf Stream is known as the Irminger Current, and it mixes with the cold Greenland Current to form a 'not so cold' current which pushes up the western side of the southern tip of Greenland. It succeeds in keeping a small part of the west coast of Greenland ice-free, so that permanent settlements are possible. These settlements are very active now that the feeding ground of the Scottish salmon has been discovered near by. On the opposite coast of Labrador there is no assistance from any part of the Gulf Stream, and almost permanent ice exists. In winter, the frozen sea stretches to below Newfoundland, where at last the Gulf Stream does play its part.

The outflows from the Arctic Ocean via the Greenland and Labrador Currents consist of cold, nearly frozen water. Sea-water is different from freshwater in its freezing behaviour, because of the salt that is dissolved in it. It is well known that a freshwater pond can freeze at the surface while the water at the bottom is several degrees warmer than freezing point. This is because the most dense water is that at 4°C., (41°F.), while of course, water freezes at 0°C., (32°F.), and forms ice, which is even lighter than the water near freezing point. In a freshwater pond, therefore, as the water is cooled it becomes more dense and sinks, until a temperature of 4°C. (41°F.) is reached. As cooling proceeds, the colder water is lighter than that at 4°C. (41°F.) and so the coldest water of all ends up at the surface, where the ice forms. Because of the salt in sea-water this 4°C. anomaly does not occur, and the colder the water gets the more dense it becomes until it freezes. Therefore cold sea-water can be really cold and near freezing point all the way down. When freezing begins to take place, pure water ice crystals appear, and these, being lighter than water, congregate at the surface. The heavier, strongly salt solution left behind sinks because it has become more dense by removal of the water. This helps to keep the deep water cold. The issue is confused along the shores of an ice-covered land mass such as Greenland by the introduction of a new supply of freshwater in the form of icebergs which break off the ends of glaciers. The glaciers have collected their water over hundreds of years in the form of snow which has compacted under its own weight and has slid gently down the slope of the mountains until its river-like life ends in the sea. The two great cold currents, both of which cling to their adjacent land masses of East Greenland and Labrador respectively carry icebergs from the Greenland fjords, and pack-ice formed by freezing at the cold sea-surface. This ice, together with another secondary effect of the Gulf Stream, provides a double hazard for shipping.

The warm Gulf Stream flowing from the tropical area of the world meets the cold Labrador Current in the vicinity of the Newfoundland Bank, an area of shallow continental shelf between 45° and 50°N, about the size of the Bay of Biscay. This area is famous for fishing and it is on the route of east-west shipping. It is also notorious for its fogs, which occur nearly fifty per cent of the time, even in summer. The fog is not caused directly by the meeting of warm and cold water streams although it is the presence of the two currents that are the prime factor in producing the fog. Just as the Gulf Stream does not directly warm the north of Scotland, so in the case of the Newfoundland Bank, the air is needed to be the carrier of heat to produce the climatic effect. Air streams blowing over the warm Gulf Stream water take up a great deal of moisture. When this water-laden air is cooled by passing over the Labrador Current, the moisture condenses in the form of fine droplets, which we call fog.

The warmth of the Gulf Stream is not needed to produce fog in summer, since warm winds from the west which have gained heat from the land and moisture from the coastal areas, are again cooled when they meet the cold Labrador stream, fed with the almost inexhaustible supply of cold Arctic water, itself augmented by the ice contribution from Greenland's glaciers. The interaction of air and water and the exchange of heat between atmosphere and sea are the important factors in the events which form the Newfoundland fogs. It is no wonder that oceanographers and meteorologists are working together to find out the numbers to feed into the calculations which are needed to explain these and other climatic phenomena. Although collection of routine data about air and water movements may seem dull, and even pointless to many who distribute research funds, these studies are essential if we are to understand the workings of the world before it is too late.

Navigation in fog is difficult and dangerous, especially in the Newfoundland Bank region, because the important shipping lines from Europe to North America converge here. Over a hundred years ago, the famous H. Matthew Fontaine Maury (see p. 8), who was one of the 'founders' of the Gulf Stream, laid down different shipping routes for east and west bound shipping. Ships travelling east were to go along the southern extremity of the Grand Banks, while the westbound traffic route was to the south of Cape Race. Shipping lines today have their chosen routes and modern navigational aids make it easier to keep to narrowly prescribed channels. Radar tells navigators when ships are in the vicinity, although occasionally the 'radar assisted collision' does take place, in much the same way that two pedestrians sometimes bump into each other, after both making several polite changes of direction.

There are some uncontrolled obstacles in the Newfoundland Bank area, which are also more hazardous on account of fog. These are the icebergs which have broken off from the glaciers in Greenland and which may measure 300 feet above sea-level at their prime. This means, of course, that some two to three thousand feet of ice is lurking beneath the surface as a root to keep the iceberg afloat. To make life more difficult for the captains of ships, the behaviour of the ice is not by any means regular from year to year; sometimes the icebergs are more numerous, sometimes they drift further south. It is not possible, therefore, to agree internationally on safe routes for shipping. The Ice Patrol watches the floating hazards each year, issuing appropriate warnings, and accidents are minimized, but the combination of fog and ice makes the approaches to Canada and the U.S.A. dangerous, even with the aid of modern reconnaissance methods. It is possible that a better knowledge of the Gulf Stream will enable the behaviour pattern of the ice to be forecast; if the forces that determine the flow of air and water are fully understood a mathematical model of the process could be defined and the outcome year by year could be calculated by computer. Until this happens, we must be prepared for the occasional accident, such as the *Titanic* disaster in 1912, when the great ship went down with 1,316 passengers and 885 crew aboard. In that instance, there was no fog; the ship was making a record first run at 22 knots and had its side ripped out by a spur of an iceberg projecting underwater, with a consequent loss of 1,513 lives. Although human beings kill each other in war and on the roads, the really big fatal catastrophes are those 'arranged' by Nature.

One of the reasons why the position of the ice is variable and difficult to forecast in the Newfoundland Bank area is the irregular boundary between the Gulf Stream and the Labrador

Current. Tongues of cold and hot water intertwine, and the melting ice may be pushed far to the south or held to a northern latitude. One of the chores of the Gulf Stream is to clean up the icebergs and the pack-ice, and it is estimated that every year about 600 cubic miles of ice must be melted. Fortunately, the Gulf Stream takes this task in its stride since it is so large, but the amount of material involved does make one realize that the balance of forces between warm and cold water is fairly delicate and could well be upset if not properly understood.

There is a great deal in common between the theory behind ocean current circulation and that of the atmosphere. Both air and water are fluids and obey the same physical laws relating to flow, and on the earth both exist on a large scale so that they are influenced by the rotation of the earth. It is interesting that the same model experiments used to find out how currents in the sea will behave are also used to demonstrate and investigate the flow of air in the atmosphere. Just as the Gulf Stream spurts out between Florida and Bimini, so there are jet streams in the air at heights flown by modern aircraft. These jet streams are important in planning flights of aircraft, just as the Gulf Stream was in sailing-ship days when Benjamin Franklin and Maury produced their advice to ships' captains. As the earth rotates so the atmosphere swirls about the earth, and breaks into eddies which form little weather patterns of their own. In a similar way, the Gulf Stream spins off pockets of energy which move off to the side of the main stream, and may in some cases produce a significant current counter to the regular direction of the Gulf Stream.

It is not known if either the pattern of air movement, or of the flow of the Gulf Stream, will ever be exactly calculable, or whether there are so many small disturbances which determine the day to day pattern, that even the most elaborate computer can never master the problem. We do know that our knowledge of air movement is improving rapidly and that one day we may be able to predict when and where pockets of incipient bad weather are building up. If we can overcome the air problem than we shall almost certainly be able to do the same with the ocean currents. It is known that the Gulf Stream changes its course and forms meanders and isolated eddies. Do these affect the climate of Europe, for example, and if so, can they be forecast? A long time ago it was suggested that the rate of flow through the Narrows of the Straits of Bimini varied, and that these variations, if fully recorded, would enable the subsequent behaviour of the current to be worked out. Variations in flow mean variations in the amount of heat transferred by the Gulf Stream from tropical to northern waters. These variations could affect the average surface water temperature, and hence the warmth carried by the westerly winds to north-west Europe. Furthermore, the changes in flow of the Gulf Stream could themselves determine to some extent the wind flow pattern and the cyclone distribution in the atmosphere. Perhaps careful observation in the Florida area of the Gulf Stream could enable forecasts to be made of mild or severe winters in Britain. The deduction will not be straightforward, however, judging by some results between 1928 and 1930. A temporary strengthening of the Gulf Stream was then noted, and while this prevailed to the south of Ireland, east winds blew, bringing cold continental European weather into Britain. As the extra pocket of warm water moved farther north, it encouraged the south-westerly winds to bathe Britain in mild winters. This is just another example of why both air and water movement must be understood if any advances are to be made, firstly in forecasting climatic changes, and possibly in the future, controlling them.

Warm air rises high into the atmosphere, and water vapour carried with the air condenses to form clouds. The clouds stop the sun's rays from reaching the earth, and so tend to control the total energy absorbed by the earth. The water which forms the clouds comes from the interaction of air and water at the sea surface. The sea in turn is stirred both by sideways-moving currents such as the Gulf Stream, and also by currents which move down to the three-mile depth of the ocean. It is interesting to note here that the ocean vertical space of about 20,000 feet is of the same order of magnitude as that which encompasses the cloud patterns above the earth's surface. Fortunately for our satellite observation both ocean deeps and atmospheric heights form only a thin skin on the 8,000 miles of the earth's diameter and therefore comparatively close orbits are all that are necessary to obtain a picture from outside of what is happening. While the atmosphere with its varying amounts of water vapour is the place where the weather action is taking place, the deep oceans, with their slow stirring and mixing due to currents, provide the area of stabilization. Heat is received at the sea surface and is stored in large packets of water such as the Gulf Stream, and is carried half-way across the earth to be given back to the atmosphere where it is sometimes beneficial and sometimes a great nuisance.

Water is a curious substance and in many ways is the backbone of our particular civilization. It is liquid over a fairly wide range of temperature, and as a liquid is capable of dissolving many different kinds of solid matter. It is most suitable for the earth, since its liquid temperature range covers the temperatures acquired by the earth when heated by the sun's rays. Water is very accommodating in freezing to a solid to form white caps at the Polar regions of the earth, and thus provide a reflecting blanket which stops too much heat from being absorbed at the earth's surface. As a second line of defence, water evaporates to form clouds of condensate in the skies, and these assist in guarding the earth from being 'cooked' by the energy radiated by the sun.

When water solidifies it expands. This is abnormal behaviour compared with most other materials, but it does mean that ice floats on water and that masses of water do not normally freeze solid because the first frozen covering of ice protects the water on which it floats, both by reflecting the sun's heat and by forming an insulating blanket. At the other end of the scale, boiling water contains energy because heat has to be applied to change water into steam. The hidden energy in water vapour, as we have seen, plays its part in the mechanisms of storms and other meteorological changes. The constituents of the water molecule—hydrogen and oxygen—are active elements which form compounds with most of the metals and other materials. Together with carbon, these two atomic partners in water, form a wide range of substances, which can burn in their simpler forms to produce energy and which, in more complicated combinations, form our life on earth. Air (which is oxygen and nitrogen), and water, together with smaller percentages of carbon dioxide, are the bases of our existence and functioning on earth but air and water are also the major controllers of the climatic habitat in which we live.

There is one particular air/sea interaction that has worried some of our scientists during the past twenty years. This is the behaviour of carbon dioxide, that gas which is produced when we animals breathe, and which plays an important part in the life cycle of plants. Just as water has many facets to its behaviour, in respect of the working of the earth, so carbon dioxide is a substance which adopts many rôles, and both carbon dioxide and water have a history which is interdependent, since, among their many and varied properties, these two substances have a mutual affinity in that carbon dioxide dissolves readily in water. One only has to look at a soda-water syphon, or open a bottle of soda water to see the large volume of gas that water can hold. The waters of the sea are a vast reservoir of carbon dioxide, and the circulation of the oceans by mechanisms such as the Gulf Stream determine the ultimate exchange of carbon dioxide between atmosphere and sea. This in turn may reflect back into changes in the world's climate. A report prepared by Dr. Roger Revelle in 1965 for the U.S. President's Science Advisory Committee said:

> Only about one two-thousandth of the atmosphere and one ten-thousandth of the ocean are carbon dioxide. Yet to living creatures, these small fractions are of vital importance. Carbon is the basic building block of organic compounds, and land plants obtain all of their carbon from atmospheric carbon dioxide. . . .
>
> Over the past several billion years, very large quantities of carbon dioxide have entered the atmosphere from volcanoes. The total amount was at least forty thousand times the quantity of carbon dioxide now present in the air. Most of it . . . was precipitated on the sea floor as limestone or dolomite. About one-fourth of the total quantity, at least ten thousand times the present atmospheric carbon dioxide, was reduced by plants to organic carbon compounds and became buried as organic matter in the sediments. A small fraction of this organic matter was transformed into the concentrated deposits we call coal, petroleum, oil shales, tar sands, or natural gas. These are the fossil fuels that power the world-wide industrial civilization of our time.
>
> Throughout most of the . . . [million or so] years of man's existence on earth, his fuels consisted of wood and other

> remains of plants which had grown only a few years before they were burned. The effect of this burning on the content of atmospheric carbon dioxide was negligible, because it only slightly speeded up the natural decay processes that continually recycle carbon from the biosphere to the atmosphere. During the last few centuries, however, man has begun to burn fossil fuels that were locked in the sedimentary rocks . . . and this combustion is measurably increasing the atmospheric carbon dioxide.
>
> The present rate of production from fossil fuel combustion is about a hundred times the . . . rate of natural removal by weathering of silicate rocks. . . . Within a few short centuries, we are returning to the air a significant part of the carbon that was slowly extracted by plants and buried in the sediments during half a billion years.
>
> Not all of this added carbon dioxide will remain in the air. Part of it will become dissolved in the ocean, and part will be taken up by the biosphere, chiefly in trees and other terrestrial plants, and in the dead plant litter called humus. The part that remains in the atmosphere may have a significant effect on climate: carbon dioxide is nearly transparent to visible light, but it is a strong absorber and back-radiator of infra-red radiation, particularly in the wave lengths from 12 to 18 microns; consequently, an increase of atmospheric carbon dioxide could act, much like the glass in a greenhouse, to raise the temperature of the lower air.

The possible 'greenhouse' effect of carbon dioxide concentrating in the atmosphere could make the earth warm up a few degrees, with dire effect for those who live near sea-level and that includes many of the most populated cities of the world. Many people, especially in this present decade of taking stock of human pollution of the earth, worry that mankind is rapidly wrecking the earth by burning fossil fuels too rapidly. The more cynical are reminded that a 200-foot rise in sea-level consequent on warming the seas, could be Nature's way of resolving the excessive increase in population that has taken place as a result of science improving medicine and agriculture. What are the dangers of our coal and oil ages putting the balance of atmosphere and oceans out of adjustment, so that the earth could become too hot to live in, or at least, a watery grave for those who live on the low-lying fertile plains?

Measurements of the carbon dioxide dissolved in the surface waters of the sea show that only one-tenth of the carbon dioxide produced by burning fossil fuels is held in this layer of water. The principal sink for carbon dioxide must be in the deep water, and in order to take up carbon dioxide from the air, the deep ocean waters must be mixed with surface water. Here again, currents like the Gulf Stream and its companion deep-water circulations play their part. One of the numbers that is not yet certain is the time taken for the circulation of the deep ocean water, and therefore the rate of the absorption of the carbon dioxide from the surface water.

The best measurements of carbon dioxide in the air suggest that about forty per cent of the carbon dioxide from burning fossil fuels is remaining in the atmosphere. However, the total amount in the atmosphere is barely showing a significant increase in atmospheric carbon dioxide to set in train a warming-up period for the earth. This may be because there is a complementary increase in atmospheric dust which reflects the sun's rays. If the carbon dioxide concentration were to rise by a factor of two, there would be almost a 3°C. (5°F.) rise in temperature. However, this is calculated assuming a temperature change does not alter the atmospheric conditions. It is probable that a small rise in temperature would produce more cloud, and this would have an automatic compensating effect by stopping some of the sun's energy from reaching the earth.

If for some reason all the carbon dioxide were to vanish from the earth's atmosphere, there would be a substantial cooling of about 10–20°C. (18–36°F.), so there is very definite concern over the carbon dioxide balance between land, air and water. The best modern assessment of the carbon dioxide situation is that the increase will not worry us in our lifetime. The rise is so slow that there will be no problem in the next four decades, even with a temporary increase in the burning of fossil fuels. The oceans act as a smoothing-out process by absorbing carbon dioxide and clouds will no doubt automatically slow down any tendency for the earth to warm up due to the small increase in carbon dioxide in the atmosphere.

There is a limit to the amount of fossil fuel in the world.

Coal and oil were deposits formed millions of years ago, and they are used at a rate measured in tens of years. It is interesting to note that the coal age, backing the steam engine, only lasted for roughly a century, from 1850 to 1950. The oil age will probably be of about the same duration, say from 1920 to 2020, when no doubt the internal combustion engine will have been phased out in favour of electric vehicles, fed by energy from nuclear power plants. These two short spans of fossil fuel power will appear to historians in a few centuries time, as man's rapid development period. The steam and internal combustion engines both allowed man to control the power of many horses. It is to be hoped that we shall not squander our fossil fuels in vain; they give us the power to move into a quiet clean nuclear age. But if we use all the fossil fuels and trip up in our advance to this nuclear age, there will be no second chance, and man will be permanently back in the horse age, from which he has so recently escaped.

5

NAVIGATIONAL PROBLEMS

'It would be wonderful', said an unknown navigator, 'if there were no currents or drifts, for then my dead reckoning would always be right.' This statement came clearly from the heart, as many ancient and modern ships' captains would testify. However, currents do exist, and the sensible navigator makes use of them rather than trying to struggle against them as did some captains of the packet boats in Benjamin Franklin's time.

Franklin collected reports from as many ships as possible that had observed the speed and position of the Gulf Stream in its course along the eastern seaboard of the U.S.A. and in its east-west run across the Atlantic. The reports were based primarily on dead reckoning observations. The ship's position was fixed every day by observing the position of sun and stars, and noting the time on a chronometer. The chronometer, invented in the eighteenth century, was a timepiece which could be relied upon to run continuously at sea, and to keep time accurately to a few seconds for several weeks. Accurate time is needed to fix positions because the earth rotates on its axis once every twenty-four hours. At the Equator the circumference of the earth covers about 24,000 miles, so an observer is flashing by the fixed stars in the sky at about 1,000 miles an hour. In order to find where he is on the ocean he must fix his earth reference point to the few seconds which correspond to a mile moved by the earth as a whole.

Prior to the invention of the chronometer by the Englishman John Harrison in 1714, navigation could be very much a case of hopeful guessing. The dead reckoning of position depends on estimates of speed (of both the ship through the water and the water itself), and in early sailing days the ship's log was inaccurate, and with slow speeds of the ship the relative effect of ocean currents was large. It is said that Columbus thought that he had sailed much farther west than the West Indies when he first made a landfall.

It is possible to determine one's position without the aid of a clock. Since the moon orbits the earth its apparent position relative to the much more distant stars varies with the lunar month. The moon then behaves like the hand of a rather slow clock, and there was a time, before Harrison perfected his chronometer, when astronomers were actively engaged in observing the moon's position, so that suitable tables could be supplied to mariners to enable them to fix their position at sea by observations of the moon and the stars. Generally, however, navigation was a very primitive matter by today's standards. Latitude, unlike longitude, could be determined without accurate knowledge of the time, so the practice of many navigators was simply to sail to the latitude of the destination and then sail as best they could east or west along the line.

Franklin collected many observations of current speed in the region of the Gulf Stream, and by plotting these on a chart was able to draw his celebrated 'river' traversing the Atlantic with speeds ranging from four down to two knots. The boundaries are today known to be drawn far too firmly, since the current changes position from time to time and at intervals sends off subsidiary swirls to the side. For these reasons, Franklin's Gulf Stream is rather wider than found in practice at any one time. However, the work of collecting observations and the arousing of interest in ocean currents was most important for the science of oceanography and navigation.

There were some experts who, on seeing Franklin's chart, believed that the sharp edge of the current could be used to fix the position of a ship at sea. The outer edge of the Gulf Stream is often clearly marked by an assortment of debris, such as old timbers which get caught at the interface of the moving waters and the stationary ocean to the west. In many parts of the Gulf Stream there is also a marked temperature difference when moving from the warm water which has come from the tropical regions to the cold waters which border the eastern coast of the U.S.A.

However, although it is often possible to recognize the edge of the Gulf Stream, we now know that the position of the Stream as drawn by Franklin is only an average and the fluctuations in course are so large that the Stream is of no use for anything other than rough estimation of position. With modern ships, whose navigation is made easy and accurate by means of broadcast time signals, it is currents such as the Gulf Stream that can themselves be studied rather than acting as useful indicators of position. Interesting observations have been made by recording the tracks of ships traversing the Gulf Stream area.

In 1950 a survey of the Gulf Stream was carried out simultaneously by a group of six ships from Woods Hole Oceanographic Institution, the U.S. Navy Hydrographic Office, the U.S. Fish and Wildlife Service and the Canadian Navy. For the first time it was possible to look at different parts of the Gulf Stream at the same moment. Although a complete picture of the current was not possible with the limited number of recording ships available, the view obtained did show the features of meandering of the main stream and the formation of large eddies at its edge. The six recording ships were spaced at roughly equal intervals along the Gulf Stream and they zigzagged across the current's path while recording pertinent data. Subsequent co-operative experiments to elucidate the behaviour of the Gulf Stream have been carried out (see Chapter 6).

The oceans are vast and do not contain easily visible fixed marks as can be found on land. Exploration of ocean features is therefore much more difficult than surveying the course of a river on land. Furthermore, as we have seen, currents like the Gulf Stream are not fixed rivers in the ocean as postulated by Franklin and Maury, and must, therefore, be looked at instantaneously if we are to obtain a true picture of their size. The problem is made more difficult in that the currents are not only at the surface, but have a variable depth, which may also change from month to month. The study of currents then, which is one of the most important researches of oceanography today—because of the effect on fish production, climate and waste disposal—is essentially one where the co-operation of many different oceanographic establishments is required. This is probably one of the reasons why oceanography is so international in character. No one nation possesses adequate ship and instrument facilities to enable it to work alone.

There is a tradition of international help and friendship on the high seas, especially in respect of scientific studies. During the fierce sea-fighting of the American War of Independence,

Captain Cook was given a free pass to conduct his voyages of discovery, and his ships were not molested by the American frigates which roamed the oceans. Today scientists from many nations meet at oceanographic conferences, and make cruises in each other's vessels, co-operating in studies of water movement, marine geology exploration and in biological work. This close communication will help to bring the different nations of the world together just as has the successful co-operative scientific effort in Antarctica.

The 1950 Gulf Stream multi-ship survey was followed by a survey based on the traditional navigational surveys using a large number of commercial ships. The Hydrographic Office of the U.S.A. organized a full year of detailed reporting from a group of oil tankers which were running from the Gulf of Mexico to the New England area and thus were travelling in the approximately north-south region of the Gulf Stream which borders the eastern U.S. seaboard.

The tankers were equipped with Loran position-fixing apparatus, and were not generally requested to alter their normal course. Most vessels prefer a course as close as possible to the axis of the Gulf Stream on their northbound passage. This was fortunate because it enabled the maximum number of readings and observations to be made. On the southbound run some vessels avoided the adverse Gulf Stream current running up the Florida coast by taking a course close to the shore, while the majority favoured an outside route. The tankers recorded their position every hour, together with sea-surface temperatures, the state of the sea and meteorological observations. As in the eighteenth century when Franklin was compiling his charts, the ships' navigators performed their dead-reckoning calculations, and these were compared with the true positions obtained from the Loran fixes. It was presumed that oil tankers, especially when loaded, would be primarily influenced by surface currents, so that the difference between dead reckoning and true position would represent the current set and drift. Since the current is capable of sweeping a ship over a hundred miles northwards in a day in the stronger parts of the Gulf Stream, the currents in this part of the world were not difficult to observe.

From the results of these experiments it was calculated that the average speeds in the Gulf Stream ranged from three knots to two knots from the south of Florida to Cape Hatteras. The speeds are less than the greatest currents that may occur in practice because of the averaging process of results from many tankers, and this process is further complicated because even one tanker on course, measuring currents over a period of a few hours, will find a net current which is lower than the maximum encountered because the ship is not always moving in the direction of the current but is sometimes cutting across a meandering side stream of the general circulation.

There was some indication of seasonal variations in the currents measured by the tankers during the ninety-four northbound and sixty-two southbound passages. In general, higher speeds by a little over one knot were recorded in July and December, but the variations depended on the particular route taken by the ships. Some evidence was obtained regarding meanders of the stream by examining in detail the observations of individual ships instead of averaging over all the ships taking a given course. The wave length, or distance between loops, of the meanders was estimated to be about 120 miles.

The Gulf Stream begins its northward run as a warm current, and loses about 1.7°C. (3°F.) by the time it reaches Cape Hatteras. The current then begins its turn to the right before crossing the Atlantic. All along the west side of the Gulf Stream, and more particularly to the north on the right-hand turn, there is a marked 'cold wall' provided by colder currents. This temperature change was used by many early observers to mark the position of the Gulf Stream, and was known as a conspicuous feature by many navigators of old sailing ships. Differences of 10°C. (18°F.) in ten miles are not unusual, and where the Gulf Stream meets the cold Labrador Current sweeping down from the north, the U.S. Coast Guard frigate *Tampa* once recorded a difference of 12°C. (22°F.) between the surface water temperature at bow and stern.

Current rips have been reported along the 'sharp' edges of the Gulf Stream for many years, and these were noted in the tanker experiments. The physical reason for current rips in the open ocean is not fully understood, but in shallower water they occur when a surface current causes small non-breaking waves to steepen up and form white breaking crests. The rips observed

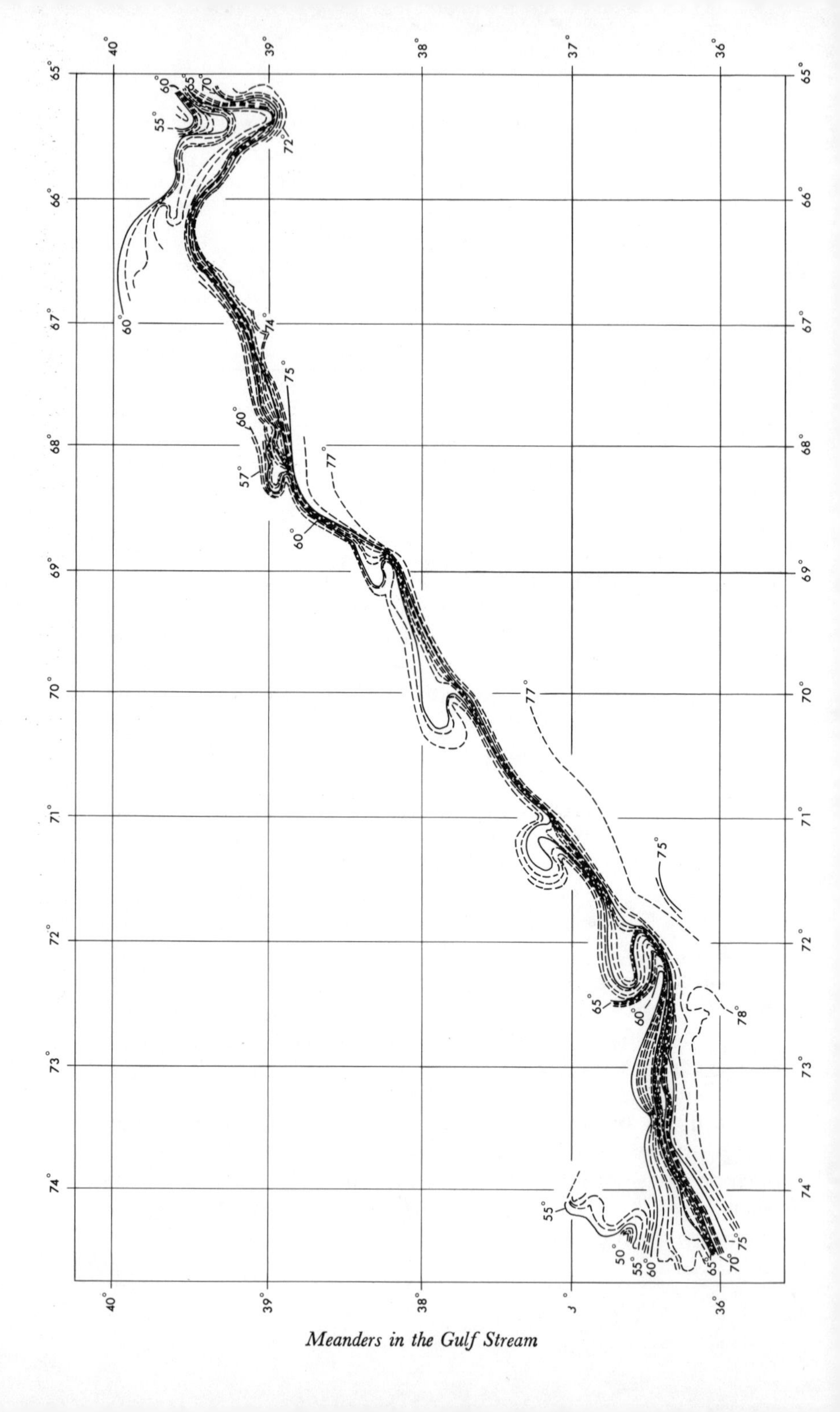

Meanders in the Gulf Stream

in the Gulf Stream experiments were concentrated around the latitude at which the current goes into a sharp bend to the east. A possible correlation between the lunar month and frequency of current rips suggests that there may be a tidal element in the effect, but the elucidation of this phenomenon is one that is still awaiting further facts.

There is a modern navigational problem concerned with currents beneath the surface. As was explained in Chapter 3, there is movement of large bodies of water at all depths in the three-mile deep oceans as well as at the surface. Now that submarines can operate at depths approximately half-way to the floor of the deep oceans, it is necessary to learn more about the deep ocean current pattern. The tools to enable us to do this, both theoretical and practical, are described in Chapter 6.

Modern submarines stay submerged for weeks at a time and positions cannot be fixed by taking star sights when 6,000 feet below surface. However, modern scientific ingenuity has invented an automatic dead-reckoner that puts surface speed and bearing measurements into the horse age. The submarine, since it provides a steady platform when gliding smoothly hundreds of feet below the surface waves, can carry sensitive instruments which record direction and acceleration. Any changes in speed or course are automatically noted by a computer which remembers everything that happened since the submarine left its base or some known fixed point. The computer calculates a dead-reckoning fix continuously from the inertial data that has been collected and could tell its position accurately if only there were no currents. We have seen, however, that the ocean water flow is not only wind driven currents at the surface, but return water streams deep down, and currents which are caused by heavy cold water or strongly saline water sinking and sliding down to the ocean floor. Just as on the surface, so deep down, currents can play havoc with the best dead reckoning. There are ways of providing fixed positions for deep submarines. The sea-bed is not flat, and contains mountain ranges and isolated peaks which, if properly surveyed and plotted on a chart, can provide landfalls which are recognizable, just as are islands at the surface of the ocean. These sea-bed features can be located with the aid of the echo-sounder and with sonar, in which a beam of sound pulses is used to search for obstacles ahead of the submarine. It is possible to place artificial aids to navigation on peaks which rise up from the ocean floor to within a few thousand feet of the surface. These can be in the form of acoustic lighthouses which send out a particular sequence of sound pulses at a specific frequency, or they can be responders which reply to a pulse sent out by the submarine.

6

AN ADVENTURE IN CLASSICAL STYLE

It could have been the precedent set by the eels taking a free ride on the Gulf Stream circuit (see Chapter 10), or merely scientific curiosity that led Jacques Piccard to drift for a month in the body of the great river. It is more likely, however, that it was just another example of the Piccard family's propensity to do something out of this world, but yet something of scientific importance with an adventurous exploring flavour to it. The twin Swiss professors were well known in the 1930s for their adventures in the upper atmosphere. Before the days of high flying aircraft, rockets or satellites, they made balloon flights up to heights of 55,000 feet (16,760 metres) to discover what changes took place in the air high up in the skies. It required only a scientific walk through the looking-glass for Professor Auguste Piccard to invent the bathyscaphe, designed to explore the ocean deeps seven miles below sea-level.

Jacques Piccard, son of Auguste, is an oceanographic engineer and worked with his father on the construction of the bathyscaphe *Trieste*. Father supplied the concept and the calculations, his son provided the engineering know-how to put these into practice. In 1957, after many years of design, trial and improvement, carried out on a shoestring budget (in contrast to the research in deep submersibles that was being done by the French Navy at the time), the *Trieste* was given financial support by the U.S. Navy on the advice of Dr. Robert Dietz (a member of the U.S. Coast and Geodetic Survey since 1968). It was decided that the craft should be the first to go to the depths of the ocean. The waters of the Atlantic, Pacific and Indian Oceans are for the most part about 15,000 to 18,000 feet (4,570 to 5,490 metres) deep, but there are some deeper trenches in the Pacific which stretch down to twice this depth—notably, the Challenger Deep, in the Marianas Trench to the south-west of Guam, discovered by the British H.M.S. *Challenger* in 1951 (see p. 70); this is the deepest part of the ocean at just short of 36,000 feet (11,000 metres). The design of the *Trieste* was such that it was capable of withstanding the great pressures at this depth, so the Challenger Deep was chosen as the location for a record *Trieste* dive. There is no great scientific reason for choosing this particular trench, but there is a certain glamour in attempting the most difficult and it must be admitted that even in the international science of oceanography there is still some friendly rivalry.

The record dive was a great success, and the *Trieste* has shown its practical value in succeeding years by diving to locate lost submarines. Piccard, meanwhile, turned his skill and experience to developing larger craft for use in shallower waters. One venture was a passenger submarine designed to take tourists below the surface of Lake Geneva. The vessel worked well, and no doubt it was a change to experience a trip on the lake taken from a fish's point of view. There was no demand, unfortunately, for similar sightseeing underwater buses in the world, so Piccard tried the research field. Several small submersibles have been made during the past fifteen years, for collecting information about the oceans, for use by the navy in research work and in recovering missiles and so on dropped in the sea, and by the oil industry for inspecting underwater pipelines and oil wells.

In 1966 the Grumman Aircraft Engineering Corporation agreed with Dr. Piccard to build a six-man craft capable of a maximum operational depth of 2,000 feet (610 metres). It was further agreed that the first venture of the finished submarine should be to drift in the body of the Gulf Stream from south to north along the North American coast. Plenty of observations had been made of the surface drift of the Gulf Stream, and current measurements and the drift of floats had shown that the current continued deep below the surface. The advantage of drifting was in the silent progress that could be made, and it was hoped that many interesting fishes would be seen, in much the same way that the crew of the raft *Kon-Tiki* saw more marine life by drifting quietly than is encountered by oceanographic ships moving under their own propulsion.

The deep ocean currents described in Chapter 3 are obviously not suitable for a Piccard type of drifting experiment. Fortunately the Gulf Stream with its flow of about 30,000 miles a year is a massive enough body of water to have strong movement at depths of 1,000 feet (305 metres) or more below surface, and is a convenient current for a team of submariners who wish to freewheel and view life underwater.

The vessel that Piccard designed for his Gulf Stream drift was built in Monthey, Switzerland, by Giovanola S.A., and shipped in May 1968 to West Palm Beach, where in August 1968 it was christened the *Ben Franklin*. The name was an auspicious one for a vessel that was to voyage in the Gulf Stream, and the story of the voyage lives up to Benjamin Franklin's character of being an enquiring, inventive and thorough philosopher.

The *Ben Franklin* is a small submersible, of about 130 tons weight, and 3,800 cubic feet volume. With a length of 48 feet, it can carry a complement of six inside its 10 feet diameter pressure hull. The design is a simple one, the main body consisting of a cylinder capped by two hemispheres. The steel is $1\frac{3}{8}$ inches thick, and extra support against external pressure is provided by a series of hollow structural rings of steel. There are two hatches, one forward and one aft, designed for safety to close under their own weight, and to seal themselves against water leaks by virtue of the external water pressure. The hatches are designed to be opened from outside if necessary in case of an emergency.

Since the *Ben Franklin* was planned to be an underwater observation vehicle, it is not surprising that it is fitted with a multiplicity of windows if such a term (rather than *port-hole*) can be used in nautical circles. There are, in fact, twenty-nine different locations for looking out into the 'inner space' of the deep oceans; they are made of plexiglass $3\frac{1}{4}$ inches thick, and provide a 90° angle of vision.

The *Ben Franklin* is powered by 376 batteries which are housed in the keel, attached underneath the pressure hull. The four 25 h.p. electric motors which drive the submarine are attached outside the main hull and for surface operation there is a Fibreglass conning-tower at the forward end. Ballast tanks are fitted at the sides of the pressure hull. When empty they provide about eighteen inches of freeboard for the *Ben Franklin*, and for diving the tanks are flooded with sea-water. The tanks can be emptied by blowing compressed air into them, the electrical control of the inlet valves of the tanks being from inside the main hull. There is an automatic safety device that blows the water out of the tanks if the submarine by mistake goes deeper than 2,200 feet (670 metres). Extra ballast tanks contain steel shot—two and a half tons in each of two tanks, as an added safety device. In an emergency a flip of a switch enables five tons of extra buoyancy to be gained by dropping all the shot. Small quantities of shot can be released by special metering valves to enable the pilot to exercise a fine control over the buoyancy of the submarine. An additional assistance in navigation is provided by fore and aft trim tanks, containing water which can be pumped from one tank to the other as required by the pilot.

The pilot can control the speeds of the four driving motors, and can also elevate or depress the motors to provide control in the vertical plane. Steering can be accomplished by use of the motors, and by two rudders acting together aft of the battery housing.

The living space, as in all submarines, is restricted, but the *Ben Franklin* is designed to be 'quite comfortable', as Dr. Piccard describes it. The two main scientific observation spaces are fore and aft, connected by a gangway some thirty-five feet long, either side of which are arranged the bunks, shower-bath, galley and bays of electronic equipment. The wardroom is in the hemispherical nose of the boat and contains a round table and seats. There are six portholes in the hemisphere, and others close to it, so that a fine all-round view is provided. Dr. Piccard had a personal porthole a foot above his pillow so that he could also see what activity was going on outside even when reclining on his bunk!

Human beings exhale carbon dioxide, and living in an enclosed space for thirty days called for some form of removal of CO_2. Panels of lithium hydroxide were strategically located throughout the vehicle and kept the air pure in a conventional manner. Oxygen to replace that used up in breathing was stored in liquid form. Various absorbers were provided to take away smells and other air impurities which might develop, while waste disposal was taken care of by chemical treatment. A neat economy, which might one day appeal to water engineers in our overpopulated areas of the world, was to use the waste water from showers and hand basins as the toilet flush. The routine on board was spartan. The expected scheduled sleep period was eight hours in twenty-four, with a few hours' rest during which the crew could nap or participate in any form of recreation. Fifteen minutes at both beginning and end of the sleep period were available for hygiene, but no dawdling in the bath was possible because water was strictly limited in supply.

Hot water was carried in insulated tanks, stored initially at 99°C. (210°F.) and cooling to 71°C. (160°F.) after thirty days. This was to save power needed to provide running hot water. Enough water, both hot and cold, was carried to provide a two-gallon shower every other day, and since one dozen changes of underwear and socks were allocated to each man there was no problem. The food was mostly freeze-dried or dehydrated, so that meals were easily prepared by adding hot or cold water. The menus were more interesting than might be imagined, including devilled ham, beef stew and chicken salad, and certainly formed a much better designed diet than the stew which appears to be the staple diet of many yachtsmen!

The submarine is equipped with echo-sounders, one of which is directed forward as an 'auxiliary obstacle avoidance capability'! The main forward-looking acoustic equipment is a sonar which can be trained 90° either side of the ship's axis and can obtain echoes from obstacles from 10 to 1,500 yards. An automatic transponder allows a surface support ship to locate the submarine. A signal transmitted from the surface ship causes a 'pinger' to send a reply signal back, so that the support ship can tell how far distant it is from the submarine. A television set is fixed on a mast above the conning tower and acts as a periscope when the *Ben Franklin* is near the surface. A second TV system is mounted just forward of the keel, and is the visual lookout when submerged. Special underwater illumination is provided by thallium iodide lights.

The Gulf Stream Drift Mission was planned to make a comprehensive set of tests and measurements for the U.S. Navy and the National Aeronautics and Space Administration, in addition to the work called for by the Piccard-Grumman organization. The submerged voyage of about 1,500 miles was to last for thirty days, starting with the initial dive off West Palm Beach, then following the Gulf Stream almost parallel to the American coast until opposite Cape Hatteras. Here the Gulf Stream leaves the vicinity of land and swings out towards the east across the Atlantic. Cape Hatteras was to be approximately the half-way point in time.

The Grumman-Piccard experiments were firstly concerned with the operation of the submarine and its performance as a scientific research vessel. In addition, a series of measurements were planned on plankton, and on mineral fluorescence in the water and on light given out by marine animals. These observations have to be made well below the surface, and preferably at night to avoid the obscuring background light from the surface.

The Navy's scientific programme was aimed at finding out more about the Gulf Stream by looking at it from a new angle—being in it instead of above it. The speed and direction of the stream and its turbulence, together with physical and sound transmission properties were required at known places and times. It was hoped that by drifting at a depth of 600 feet (185 metres) a direct measurement of current speed could be obtained to confirm the estimates made in the past. Apparatus was mounted on deck to measure temperature, salinity, sound velocity and water pressure, as well as observations of the current speed and direction calculated from the position of the *Ben Franklin* as determined by the surface support ships.

It was hoped that the composition of the deep scattering layer would be investigated as it rose and descended past the submarine. This layer was noticed originally as 'false echoes' produced on echo-sounder records of surface ships, and is now known to be a layer of concentrated small animal life which rises to the surface at night and descends to several hundred feet during the day. In the early days of echo-sounding the deep scattering layer accounted for several reports of shoals which on careful investigation by survey ships proved to be non-existent. The reason for the movement of the layer is that the animals of which it is composed are sensitive to light, and they follow the level of light that they prefer, moving down away from the bright daylight and up to get the most they can during the night. The naval scientists were equipped with acoustic transmitters and with cameras to learn more about the detail of the deep scattering phenomenon, and combined these observations with measurements of light intensity.

Experiments were planned to study the sea-bed, and for this

purpose six excursions to depths of 2,000 feet (610 metres) were arranged so that the sea-floor could be examined and photographed. A sideways-facing sonar device was used to record the profile of the sea-floor 200 feet either side of the submarine, while the vessel was thirty feet from the bottom. A sub-bottom profiler, which is a type of echo-sounder whose signal can penetrate the soft sediment, was employed to determine the shape of the ocean floor bed-rock, and measurements of the loss of signal strength on reflection at the sea-floor were made to provide more information about the behaviour of acoustic devices at sea. The support ship fired small explosive charges as the sound source for some of these experiments. In addition to the direct geological observations, gravity and magnetic measurements were planned, using an automatically recording Lacoste-Romberg gravity meter and a magnetometer. Gravity measurements at sea in the early days of the 1930s were always made in submarines because these vessels provided a stable platform for the gravity measurements which must be made to better than one part in a million and are very sensitive to roll and pitch. It is possible to observe gravity from a surface ship using newly designed stabilized platforms, but a drifting submarine with stabilized apparatus should provide the best possible measuring conditions.

Small groups of people living in confined quarters face special problems of living, and these are of great interest to those who travel in space. For this reason NASA proposed to use the *Ben Franklin* to observe the operational efficiency of the people involved during their thirty-day confinement aboard. NASA provided an engineer, Chester May, to accompany the expedition, and to use his five fellow-members of the *Ben Franklin* as guinea-pigs. Three cameras placed in strategic positions recorded the functions of all on board every two minutes for the duration of the trip, while under each mattress a counter was placed to log the resting time taken by everybody. Counters on the floor recorded steps and at regular intervals Chet May took skin from his human sampling points, and swabs from the sink, port-holes and the floor so that colonies of bacteria could be cultivated to ascertain what microbes were multiplying on board. Analyses of the air were made at frequent intervals, not only to monitor for dangerous concentrations of oxygen or carbon dioxide, but also to see what else in the way of contaminants would appear.

Two surface support ships were in attendance during the thirty-day voyage. The job of the M.V. *Privateer* was to track the *Ben Franklin* with the help of the transponder aboard the submarine, together with accurate position fixing of the M.V. *Privateer* herself. Surface measurements of temperature, salinity and chemical composition of the sea-water were also part of the support ship's programme. The U.S. Navy ship *Lynch* towed a special trawl designed to collect samples from the deep layer, and fired the charges for the bottom reflection experiments made by the *Ben Franklin.* An aircraft fitted with airborne radiation thermometers, traversed the Gulf Stream to determine the Stream's limits by means of the temperature difference between the Gulf Stream and neighbouring ocean water. This information allowed scientists on board the *Privateer* to calculate the submarine's position relative to the centre of the Gulf Stream current.

In addition to Dr. Piccard himself, and Chester May (the NASA engineer) there were four other members of this notable expedition. The Captain, Donald J. Kazimir, was a former U.S. Navy submariner and was then working for the Grumman Corporation. The pilot, Erwin Aebersold, was a Swiss who had worked seven years for Dr. Piccard; he is an experienced air pilot and a precision-minded technologist. Two oceanographers completed the ship's crew: Frank Busby, a graduate of Texas A and M University was a civilian employee of the United States Naval Oceanographic Office, and probably knows more about submersibles than any other living man; Kenneth Haigh was from Britain, on exchange at the time from the British Navy's scientific service to the U.S. Navy; he is an expert in echo-sounding, sonar and all underwater acoustic measurements.

Communication between *Ben Franklin* and the M.V. *Privateer* was maintained by an underwater telephone system which had a range of 7,500 yards (6,860 metres). Another indirect communication link was provided by an apparatus on board the *Ben Franklin* by means of which plastic spheres approximately five inches in diameter could be shot out of the submarine so that they floated to the surface. These spheres could be used to send water samples, rolls of film, recorder paper rolls or written messages to the surface. The communication by this system was, of course, one way only.

The cruise of the *Ben Franklin* in the Gulf Stream was a resounding success and is a tribute to excellent design and planning. Everything worked so efficiently that in this day of exaggeration and catastrophic headlines, the operation did not receive the publicity it well deserved. One of the disappointing aspects of the drift was the lack of a concentrated deep scattering layer. It is possible that the marine life that forms the scattering layer finds the Gulf Stream itself somewhat too warm; animals that are so sensitive about their light conditions could be fastidious in other directions. A small number of fish were observed; for example, Dr. Piccard says: 'once a small squid came and attached itself to the windowsill above my bunk, and we observed each other in complete tranquillity'. Frank Busby reported that a two-inch lantern fish followed the *Ben Franklin* right from West Palm Beach to Cape Hatteras, where it was eaten by a ten-inch fish, and as Busby remarked, 'there went our deep scattering layer'. A school of tuna were seen one day, and on blowing some air from the submarine the tuna played in the bubbles as if taking a bath. They were presumably rationed for air just as the *Ben Franklin*'s crew were for water! One barracuda, two swordfish and a few sharks were also observed.

Some of the highlights of this pioneering voyage may be gathered from excerpts from Dr. Piccard's diary:*

> July 16th. All during the night the Franklin has drifted slowly at about 600 feet . . . we are at a point some 60 miles southeast of Cape Kennedy. We send a message to the Apollo 11 crew, a few hours before they leave for the moon. At 9.32 a.m. we hear—indirectly by way of radio and underwater telephone, the countdown and departure of the most fantastic expedition ever undertaken by man.

> July 18th. At the beginning of our fourth day, as usual, we receive our position as determined by the Privateer. We mark the point on the chart. We count the miles. Our speed

*As reported in Dr. Piccard's article in the *New York Times* of Wednesday, 20 August 1969.

is good. But wait. Suddenly Frank who has marked the chart, and who also keeps an eye on his electrical thermometer giving the temperature of the outside water, accurate to two-hundredths, exclaims, 'They're wrong, or did we misunderstand them? We cannot be where they say we are. The temperature does not fit'. We ask for our position again, and there is a correction. Frank is right. Navigation by thermometer is a reality.

We prepare tonight for a new descent to the bottom. Frank declares, '450 metres, the temperature will be 7.15°C'. At times like this we ask ourselves, why is he here, when he knows all the answers already. But it was just for that reason that I chose the Gulf Stream and not some unknown current for study. When the base is relatively well established, every new bit of data can be integrated into the system and become useful.

At 8 o'clock I open the valve of our variable ballast tank for descending. Every pound of water taken in will have to be blown out later on and we must spare our compressed air. At 8.30 we arrive only 30 feet above the sea floor. We are well equilibriated and well oriented because of the guide rope that is suspended from our craft.*
On the sea floor are a few crabs four to six inches in diameter.

A few fish, a sea anemone slowly moving its tentacles, a small fish darting away from the claws of a crab—all this we witness from our floating observation station.

July 19th. The assault occurred at 6.09 at 252 metres down.

As a matter of fact it was really an attack; short, precise. The swordfish was about five or six feet long. Another one was waiting for him at the limit of our visibility. The combatant rushed forward and apparently tried to hit our porthole, missing it by a few inches. Then he circled around for several minutes close to the boat. Content that his domination of this position of his realm was not threatened, he joined his friend and left, never to be seen again.

This attack is extremely strange, because last year another

*A weighted rope was hung below the *Ben Franklin* so that they received due warning on approaching the sea-bed.

swordfish attacked another submersible, the Alvin. We were probably considered an underwater monster, and our portholes were thought to be eyes. But why, I wonder, such an attack, and what courage for such a fish to take on a 130 ton submarine.

July 20th. During the day we wait with impatience for the news of the moon landing. The message arrives finally at 4.20 p.m. and is short and precise without any comments. 'Two Americans have landed on the moon'.

August 11th. The day after tomorrow at 8.30 p.m. we will have completed a full 30 days adrift. Over five million measurements of temperature, salinity, speed of sound and depth have been recorded. About 1000 explosions with their multiple echoes have been recorded on magnetic tapes. The contents of the seawater, principally its chlorophyll and various minerals, have been measured regularly several times a day. The earth's gravity has also been measured during a total of 24 hours.

August 15th. On board the Coast Guard cutter Cook Inlet. The dive ended yesterday at 8.0 a.m. without incident.

We can also state categorically that we have unlocked more questions about the Gulf Stream than our journey answered.

These quotations from Dr. Piccard's diaries emphasize the smoothness with which this well planned expedition was run. The vast amount of data that various scientists and organizations desired was collected steadily and accurately. There is one obvious advantage of making oceanographic measurements from a submarine rather than from a surface ship; the craft is always steady because the sea is calm. Work does not have to stop because of bad weather; apparatus does not get damaged by waves or by rolling of the ship, and of course no one feels seasick. Sometimes the *Ben Franklin* did experience quite sudden changes of depth when drifting with the Gulf Stream. With the submarine at steady trim, a depth change of sixty feet (eighteen metres) in a period of ten minutes would occasionally take place. This could have been due to internal waves in the deep water. These occur at the interface between two layers of water of different density, in much the same way that waves occur at the sea surface, which is the interface between sea and air.

The highest drift speed was close to five knots, and the whole of the voyage was much faster than had been forecast from previous knowledge. A maximum speed of 2.7 knots had been expected. At one stage of the journey the *Ben Franklin* drifted into a large eddy, which was being whirled off the main stream. It was necessary for the support ship to tow the submarine back into the Gulf Stream, but the continuity of the experiment was not broken, since the hatches were not opened. The drift down the Gulf Stream is similar to the descent of the *Trieste* to the deepest valley of the oceans, or the visits of the Apollo astronauts to the moon. Measurements from a distance, either at the sea surface in the cases of the Gulf Stream or the Challenger Deep, or by telescope and satellite photograph for the moon, make it possible to deduce, usually correctly, what is to be found. But we are never quite certain until we as human beings go and see for ourselves. We now know that the Gulf Stream does flow hundreds of feet below the surface, that it is a fast-moving body of warm water, and that it meanders and eddies in the course of its progress around the Atlantic. We know these things for sure because Dr. Piccard and his team have been there.

7

HYDROGRAPHY TODAY

In order to carry out effective naval and military operations it is essential to be able to traverse the ocean swiftly and accurately in order to land troops and stores at the right place and time. While the last remaining continents were being discovered in the eighteenth century, the art of drawing accurate maps was progressing rapidly. Charts are needed by navigators to allow them to guide their ships to their destinations in safety. Therefore, in addition to the shape of the coastline, which cartographers who make land maps depict, the chart-maker must provide additional information. This extra help to the navigator takes the form of depth observations adjacent to the coast, so that navigable channels may be followed, and of oceanographic information, especially that relating to currents and to hazards such as tide-rips and moving sandbanks, and the nature of the sea-bed.

British hydrography stems from the great voyages of discovery led by Captain Cook in the latter part of the eighteenth century. In his book *The Admiralty Chart*, which gives the history of British naval hydrography in the nineteenth century, Admiral Ritchie points out that Cook was the main influence on the great line of British hydrographers in the nineteenth century. Although the British Navy entered the chart-making field nearly a century after the French, the excellent standards and the inspirations to hard work in distant parts of the world soon put the Admiralty Chart in the forefront of those used by navigators of all nations. It is worth quoting a paragraph from Admiral Ritchie's book to remind one of the famous names of ocean surveyors who served with Captain Cook and learnt their skills from him.

> The true greatness of Captain Cook has become clear to me whilst writing this book, for his influence runs on through the nineteenth century. Whilst developing his impeccable running surveys which have stood the test of time, he formed

a school of surveyors among the junior officers who sailed with him to the Pacific. Men like Vancouver, Riou and Bligh passed on their knowledge in turn to Mudge, Broughton, Flinders and others, and thus we can trace through many lines the handing down of knowledge to Richards, Evans and Wharton at the end of the century, the original techniques, constantly improved upon, with the dedication to accuracy unimpaired. It is not surprising that Admiral Wharton, the last Hydrographer of our story, edited and published the Journals of Captain Cook.

With the traversing of the southern land mass of Antarctica by Sir Vivian Fuchs, followed by the multi-nation attack on geographical and geological evaluation of this last continent, there are no new land areas of the world to conquer. The hydrographic services of the world are still active, however, since although aerial photography has speeded up land surveys, it is still necessary to survey the approaches to land and to make more detailed charts of coastlines which are being pressed into use to make new harbours to serve our rapidly developing and increasing world population.

We are in the middle of the oil age of man, and although one day we shall probably be finished with the rather noisy and sometimes smelly combustion engine, oil and its products have proved on balance to be of great material benefit to the majority of human beings. Every year hundreds of millions of tons of oil are moved from the producing areas, which by some odd quirk of geology are generally in sparsely inhabited parts of the world, to the centres of population where petroleum products are used. It has become obvious during the past twenty years that economies in transport are most readily obtained by employing large ships. Whereas the 10,000 ton tanker was a very large vessel in pre-war years, it is now dwarfed by the giant 350,000 tonners used by Gulf Oil Company on a shuttle service from Kuwait to Europe. In order to operate these large ships safely, it is necessary to locate rocks and shoals in much deeper water than in the past, so that careful chart-making of deep water approaches has provided a new field of activity for the hydrographers of the world. The largest tankers cannot now negotiate the Suez Canal (even if open) and must make the voyage around the tip of South Africa; they cannot even travel up the English Channel, but must find deep-water ports such as Bantry Bay in Ireland or Milford Haven in south Wales.

The world's demand for oil has been doubling every ten years since 1945, and to cope with this loading terminals at the oilfield end of the journey, and jetties to receive the oil at the consumer end have been constructed. There is another form of oil activity which has grown up during the past few decades. It is estimated that one quarter of the world's oil lies beneath the sea-bed on the shallow water continental shelf areas of the world. Already 17 per cent of our oil is being produced from wells drilled in shallow seas. The activity associated with offshore exploration and production has provided plenty of survey work in many parts of the world, and the tempo of this activity is increasing as oil companies turn to many new continental shelf prospects. The construction of drilling platforms, and all the apparatus needed for production, together with loading terminals a long way offshore to cater for the deep draught giant tankers, has revealed a new side to hydrographic studies. The nature of the sea-bed itself, since it affects the driving of supporting piles, or the burial of pipelines on the sea-bed, is as important in planning as the shape of the sea-floor. When choosing a site for a loading terminal it is necessary to consider the currents in the sea, together with the prevailing winds and the likelihood of rough weather. Hydrography is, in fact, expanding to cover the wider aspects of oceanography. It is a sign of the times that, a few years ago, the U.S. Hydrographer changed his title to that of Oceanographer of the U.S. Navy. As was so aptly said at the time: 'It is now important to look at what's in the bucket as well as the shape of the bucket itself.'

Since 1945 the navies of the world have been looking at hydrography and oceanography of the deeper parts of the bucket. Previously, with a commitment to make charts adequate to guide surface ships to a safe anchorage, there was not much interest in naval circles in the sea-bed beyond the 100-fathom (600 feet) line. Now that modern submarines travel for long periods of time at depths of thousands of feet, the potential obstacles in deep water must be mapped. Thanks to this new interest, our knowledge of the shape of the oceans has increased rapidly during the last few years, and the elegant and expressive charts produced by Dr. Heezen and his co-workers at the Lamont Geological Observatory, Columbia University, show these discoveries in a pictorial form which has certainly assisted in formulating some of our new theories of the geology of the ocean floor. These maps of the sea-floor, for this is what they are, rather than charts with soundings and water depth contours, enable geologists to compare the sea-bed topography with the topography depicted on land maps, and thus to view the rocks of the earth's surface as a whole, as if the water had been removed from the oceans to reveal the original earth form.

It is much easier now to map the ocean floor than it was in the nineteenth century, when the first surveys of the deep sea were made. These early soundings were made for curiosity in the first instance, but soon demonstrated their value in selecting routes for trans-oceanic telegraph cables. The depth of water in the old days was measured by heaving the lead—that is, lowering a heavy sinker to the sea-bed and noting how much rope was paid out when the lead came to rest. When the water is three miles deep, as it is in many parts of the oceans, this process is most time-consuming. A specially lightweight grass rope must be used, otherwise the weight of miles of rope hanging over the side would mask the change in suspended load when the weight finally reached the bottom. Having lowered the weight to the bottom, a team of seamen had to apply themselves to the windlass to haul it back again. In the 1872 *Challenger* expedition, which made the first comprehensive set of deep ocean soundings all over the world, only about 250 deep soundings were made in the course of a three-year voyage. These formed interesting spot samples of depth, however, and one of them was by chance within a few miles of the deepest part of the ocean. This is in a valley called the Marianas Trench in the south-west Pacific Ocean. Here the water is about 36,000 feet deep. By coincidence this greatest ocean depth was discovered by a British Navy ship, also named *Challenger*, and the deepest spot in the oceans is called Challenger Deep. Even in 1951, when modern echo-sounder depth recorders were available, a check depth determination was made by the officers and scientists on board, by lowering a lump of pig-iron on the

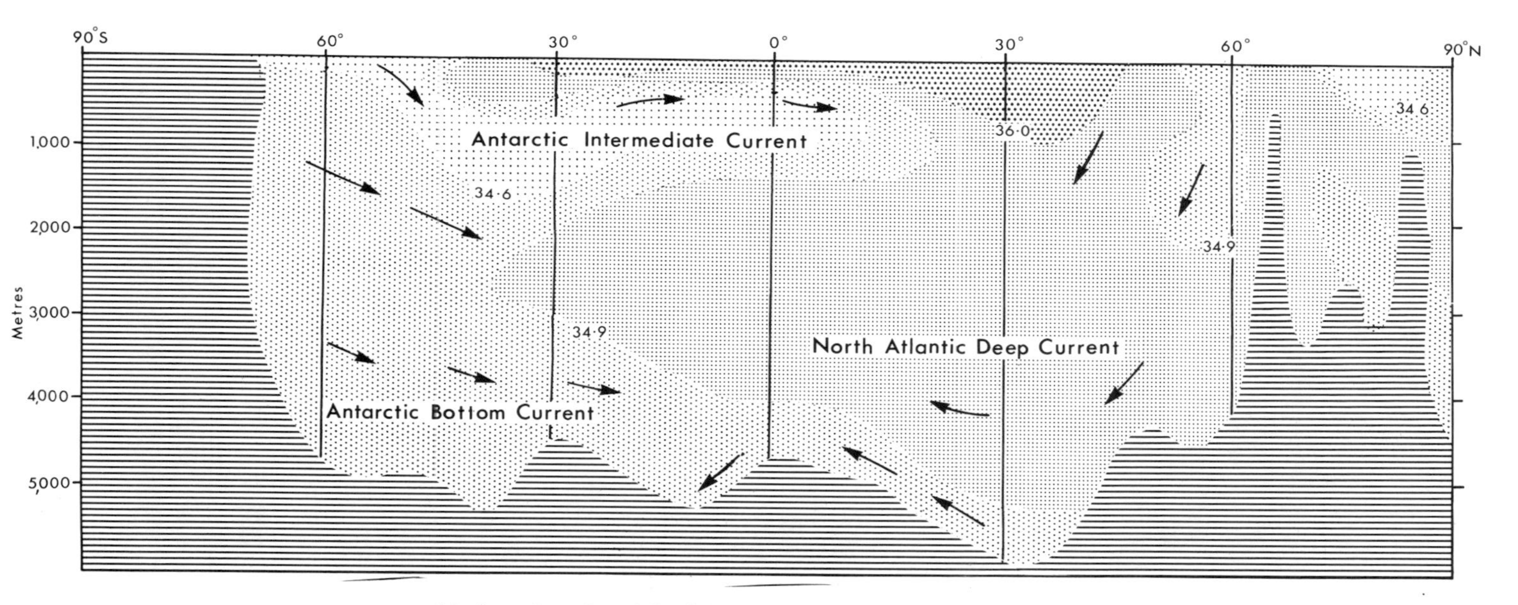

North–south section of the Atlantic showing different levels of water

end of fine steel piano wire. This was somewhat less laborious than the nineteenth century method, and they did not bother to pull the weight back to the surface.

The old rope or later wire soundings both suffered from the disadvantage that they took several hours to make, and during that time wind and surface current may have drifted the ship a considerable distance. Furthermore, the fixing of the ship's position was not too exact in those days. For these reasons, it was not possible to draw the shape of the underwater features such as mountain ranges, flat topped sea-mounts or underwater volcanic peaks. In places such as the Gulf Stream, where the currents are swift, position-fixing makes investigation very difficult, although, as we shall see in Chapter 8, in 1888–9 Lieutenant Pillsbury and the *Blake* managed to fix their position by anchoring in the mile or so depth of water in the fast-moving part of the Gulf Stream.

The modern echo-sounder measures the depth every few seconds, by emitting pulses of sound which travel to the sea-bed and reflect back to the surface at a speed of a mile a second. It is thus possible to steam along on the surface and draw profiles of the shape of the sea-bed continuously. Modern aids to navigation such as Decca or Loran equipment have assisted in placing these profiles in their true geographical position, and the latest satellite navigation system gives a greatly improved accuracy. It is still not possible to cover large areas at one view, as happens on land when surveys are made from aircraft with the help of aerial photography. However, by traversing back and forth, the detailed shape of deep sea features has been discovered, allowing our present extensive, though by no means complete knowledge of the sea-bed shape to be gathered. There is a new device which may help to give a broader sweep to our deep-sea surveys. This is the British National Institute of Oceanography's 'sideways echo sounder', named 'Gloria'. This is a large towed body which carries a powerful sound source, whose beam of energy points sideways, effectively illuminating a wide swathe of the sea-bed with its echo-sounding pulses. As the parent ship tows Gloria along, a recorder draws a picture of the heights and depressions of the sea-floor which lie to one side of the ship's course. Future developments that are envisaged may make it possible to take a form of aerial photograph from the sea surface, using powerful sound sources in place of light rays.

Now that oceanography rather than hydrography has become the work of the navies of the world, routine measurements of the variations in the earth's gravitational and magnetic fields are made from hydrographic ships. These measurements tell us a great deal about the geological structure of the sea-bed, and help to build up a picture of the earth, and the changes that have taken place in the 4,500 million years of its existence. In addition to making measurements which help the earth scientist, the world's navies are continually adding to the store of knowledge concerning currents, not only in the Gulf Stream, which is by no means fully understood, but also in the other surface currents of the oceans, and in those hidden deep ones which affect the operation of submarines.

The shape and depth of the sea-bed plays some part in deciding the course of the Gulf Stream. just as the existence of a land barrier at the west side of the Atlantic Ocean determines the initial deflection of the Equatorial surface currents that flow into the Caribbean before being deflected through the Straits of Bimini to make the Gulf Stream jet (see Chapter 3). In addition to this type of basic information, hydrographic ships make many direct current measurements, and, perhaps more important, they make indirect observations which allow the speed and path of both surface and deep currents to be calculated and deduced. These indirect observations are somewhat tedious and unexciting, and are rather scornfully looked down on by marine geologists and geophysicists, whose experiments yield results of more immediate interest.

'Water bottling', as measurement of temperature and salinity of sea-water is called, consists of taking a sample of sea-water at a known depth, and recording the water temperature of that depth. A series of 'reversing' bottles is fixed to a cable which is lowered over the side of the oceanographic ship. When the vertical line of bottles is at the predetermined depth a weight or 'messenger' is slid down the cable. When the weight hits the top bottle of the line a catch is released and the bottle turns over and a top and bottom flap close to seal off a sample of sea-water. The point of reversing the bottle is to fix the record of temperature which has been reached by a thermometer attached to the bottle, the thermometers being so designed that they record normally when upright, but lock the position of the mercury when upside down. The triggering of the first bottle releases another messenger which travels down the cable to trip the second bottle. The process continues down the wire until the series of a dozen or more bottles have all secured their water samples and have reversed to freeze their temperature recordings. It only remains to haul the cable to the surface, read the temperatures, and take the water samples to the laboratory for analysis.

As was hinted in earlier chapters, the temperature and salinity measurements perform two functions. In the first instance they allow large bodies of moving water in the oceans to be followed by means of their conspicuous salinity and temperature levels. Thus the Gulf Stream waters are warm from their Equatorial origin, and have a salt content different from, say, those brought down from the Arctic by the Labrador Current. It is possible to determine whether the Gulf Stream is confined to a few hundred feet of surface sea, or whether it extends to the sea-bed, and it is possible to see which currents are tributaries of a main stream, and which have an original entity.

The second important result of temperature and salinity measurements is the calculation of pressure due to the column of salt water that exists at different depths and at different places in the oceans. These pressures, like the air pressures in the atmosphere, determine in which direction the water will flow—just as winds in the atmosphere blow from high to low pressure areas, so ocean currents flow from regions of heavy cold salt water to those where the density is less. The problem for the ocean current measurer is much more difficult than for the meteorologist, because of the scarcity of observing stations. The atmospheric pressure is recorded continuously at many places distributed over the earth's surface, so that at any one moment a synoptic picture of the atmospheric pressure can be drawn. It is impossible to cover the sea with enough ships to obtain continuously a detailed 'flash' picture of ocean conditions. Measurements of the Gulf Stream are made by a few ships over the course of many years, and gradually a pattern of behaviour is worked out, but care must be taken not to fall into the error made by Franklin, by plotting all observations as if the course of the Gulf Stream was fixed, as is that of a river by its confining banks.

The study of ocean currents is one in which it is vitally important that as many nations as possible shall co-operate, so that the maximum number of observing ships can be brought to bear on a given area. It is only in this way that true transient effects of a current can be distinguished from the main flow.

During the Second World War a simple temperature-measuring apparatus was invented by Professor Athelstan Spilhaus, an American scientist. The bathythermograph is a heavy cylindrical body about three feet long and a few inches in diameter which is attached to a recovery wire and then thrown over the side of a ship, which can be a destroyer moving at speed. The bathythermograph contains a pressure element, which effectively registers the depth at any instant of fall, and a quick acting thermo-couple element which records temperature. These two active elements move a stylus across a glass plate, and thus draw a graph of temperature against depth. The bathythermograph records temperatures down to a few hundred feet, and since many naval ships use the apparatus, provides a vast series of water temperature measurements over all the sea surfaces of the world. In particular, these measurements show the depth at which the top waters of the sea, warmed by the sun's rays, meet the cooler ocean water beneath. The results are useful in studying such currents as the Gulf Stream, but were made in wartime for a slightly different purpose. The sonar device, which is used for detecting submarines, is virtually an echo-sounder which sends its beam of pulses out ahead of the ship rather than downwards. If a submarine moves across the beam of sound, reflections from the metal hull are sent back to the recording part of the sonar on board the searching ship. However, sound rays do not behave regularly when they encounter layers of water of different temperature. The sound tends to bounce off a layer of cold water and, therefore, could miss a submarine hiding below the junction of warm surface and cold deep waters. A knowledge of the positions of this interface is not only useful to the attacker in warning him of possible 'blindness' of his sonar, but it also enables submarine captains to evade the sonar beam. In more recent years layers of warm water at depth have been used to contain sound waves which

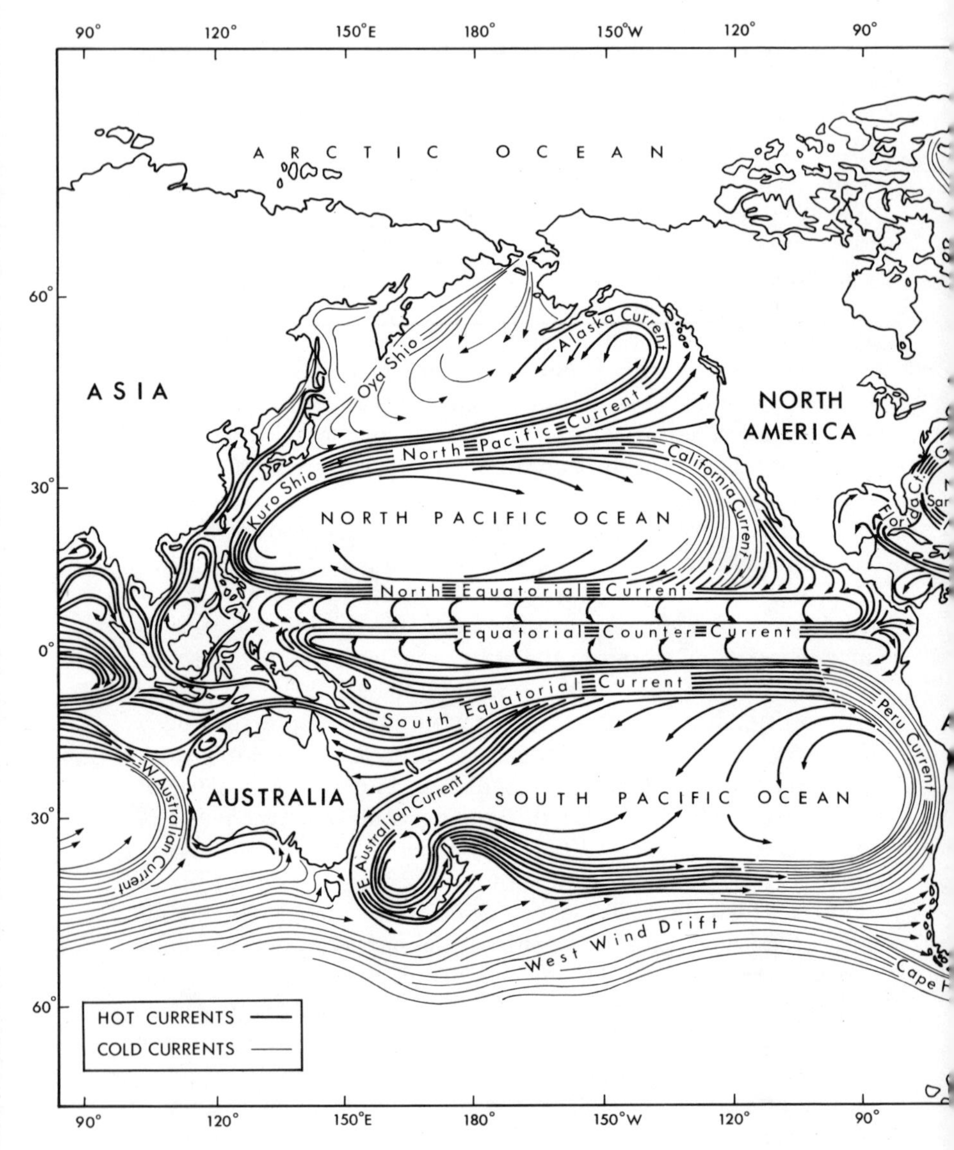

The world surface currents

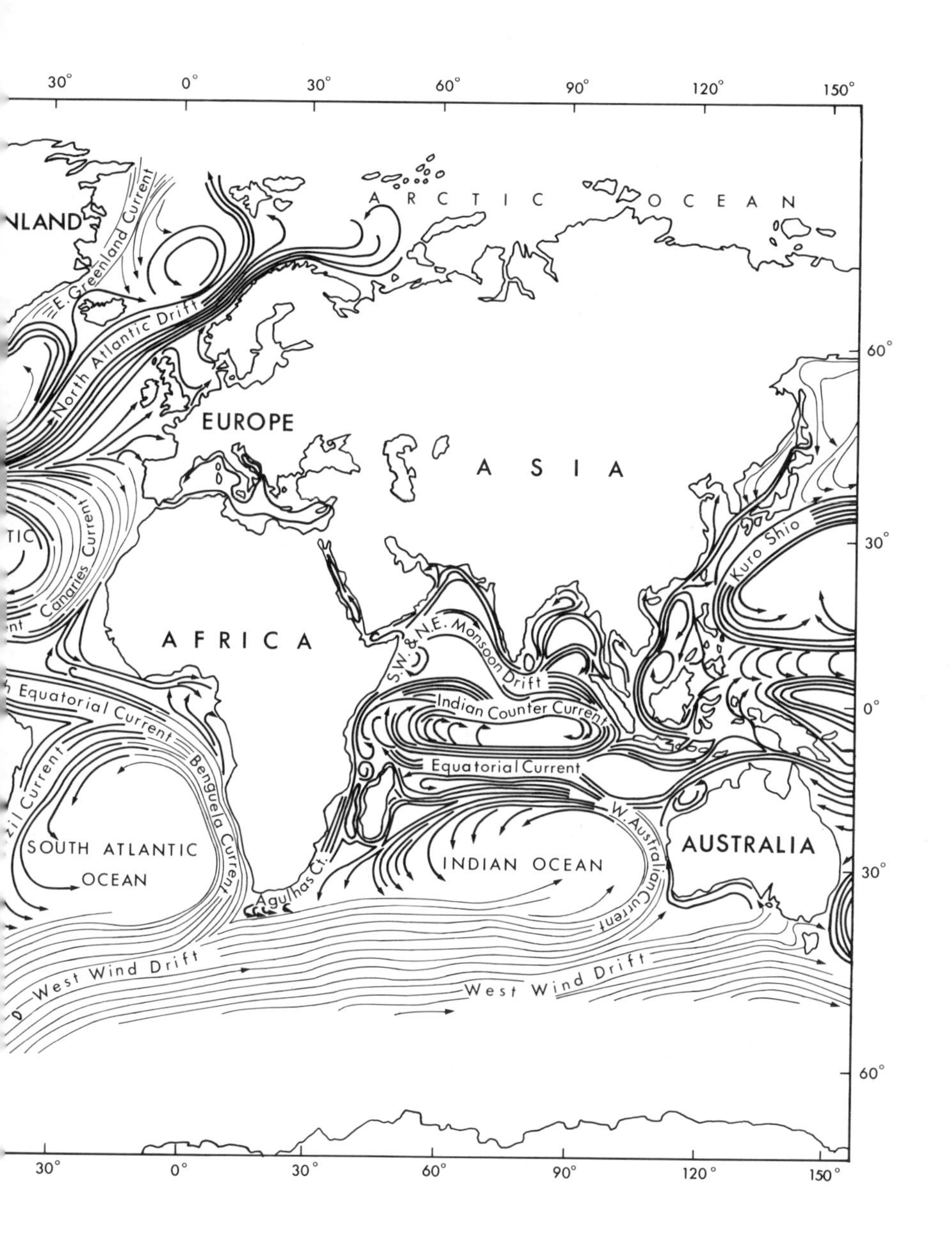

30°
0°
30°
60°
90°
120°
150°
ARCTIC OCEAN
NLAND
E. Greenland Current
North Atlantic Drift
EUROPE
ASIA
TIC
Canaries Current
AFRICA
S.W. & N.E. Monsoon Drift
Kuro Shio
Equatorial Current
Indian Counter Current
Equatorial Current
Benguela Current
W. Australian Current
AUSTRALIA
SOUTH ATLANTIC OCEAN
INDIAN OCEAN
Agulhas Ct.
West Wind Drift
West Wind Drift
60°
30°
0°
30°
60°

are bounced from top to bottom of the channel and thus can be propagated for very long distances for signalling purposes. In one interesting case, a volcanic explosion, which was probably the cause of the disappearance of a Japanese oceanographic ship while investigating an underwater eruption, was recorded thousands of miles away, the sound signals being channelled by a low-density water layer.

The direct current measurements, and the currents deduced theoretically from observations of salinity and temperature, have enabled us to locate a whole pattern of currents in all the oceans of the world. While the general form of the currents is similar, as would be expected since they are initiated by the world atmospheric wind system, and are influenced by the rotation of the earth, each ocean does have its idiosyncrasies. The differences between one ocean and another are largely determined by the land obstacles, together with the unseen sea-bed irregularities, which distinguish the different parts of the globe.

In the Pacific Ocean there is a direct counterpart of the Gulf Stream, the powerful Kuro Shio which runs along the eastern coast of Japan, and turns eastwards to circulate in a clockwise direction around the ocean. In the southern hemispheres of the three great oceans, Atlantic, Indian and Pacific, there is a surface current pattern which is the mirror image of that in the northern oceans. Since there is no land to stop it, a circulation sweeping eastwards joins up the anti-clockwise current patterns at the lower part of the three southern oceans. There cannot be a northern hemisphere counterpart of this current because the continents, which are distributed unevenly between north and south hemispheres, obstruct the flow of water.

The water in the Equatorial regions of the Pacific Ocean is driven westwards by similar wind forces which start the Gulf Stream movement in the Atlantic. The Pacific North Equatorial Current lives always in the northern hemisphere, while the South Equatorial Current ranges between the Equator and 4°N. The reason for this asymmetry is presumably due to some displacement from the central Equatorial band of the earth due to the uneven distribution of land on the earth's surface. There is an Equatorial Counter-Current, situated just north of the Equator and running between the main wind-driven Equatorial

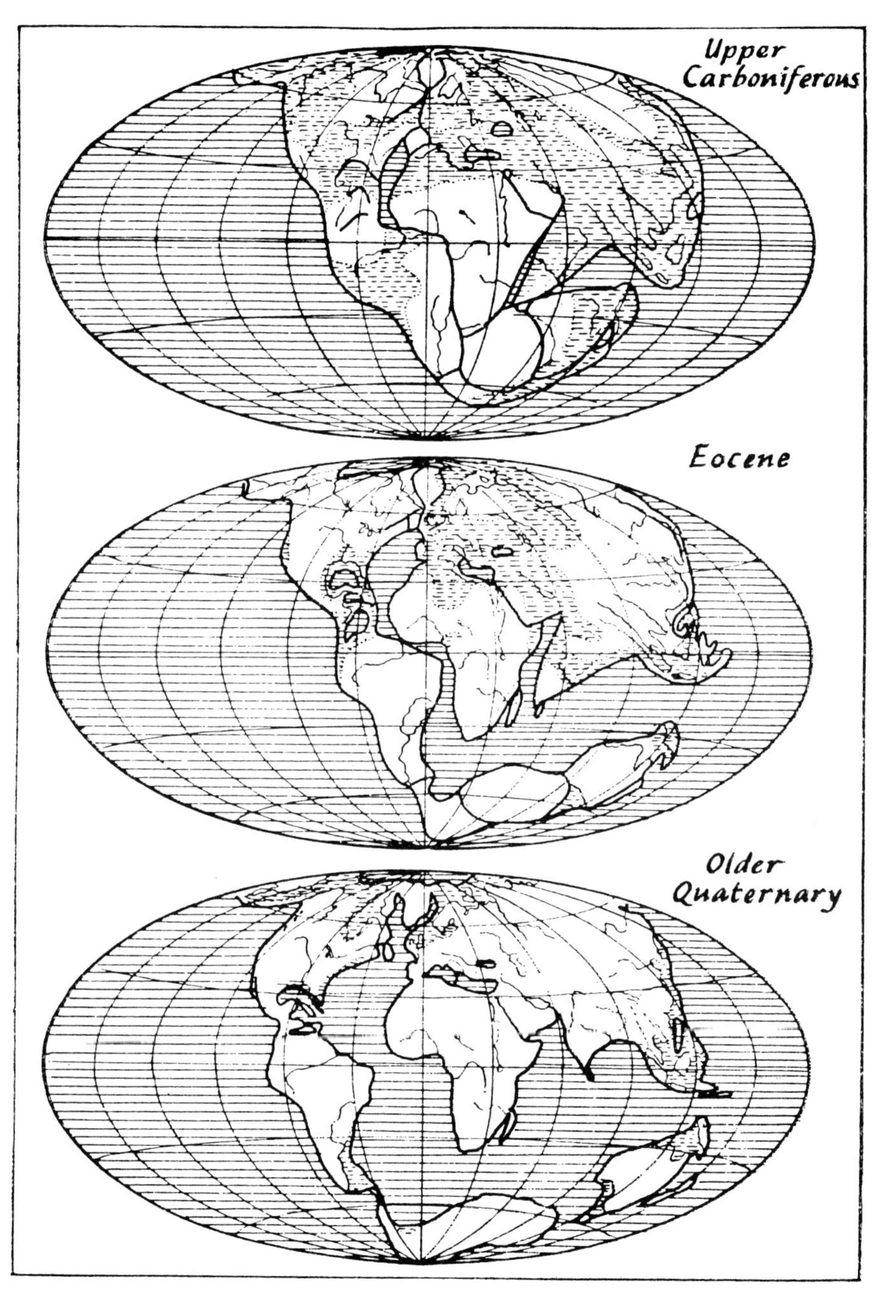

How the world looked according to Wegener about 250 million years ago (upper), *60 million years ago* (middle) *and one million years ago* (lower). *The light shaded areas are shallow seas.*

Christopher Columbus, from an engraving by Montanus (Mansell Collection)

Above: *drawing of 'Bing' from the* Illustrated London News *of 1848.*
Below: *Sea Monsters drawn from a chart of 1585.*

Above : *the submersible,* Ben Franklin, *capable of operating at depths of up to 2,000 feet.*

Below : *the interior of the* Ben Franklin *showing the hot-water tanks and kitchen.*

Right : *the pilot console of the* Ben Franklin.

Benjamin Franklin's chart of the Gulf Stream, 1770.

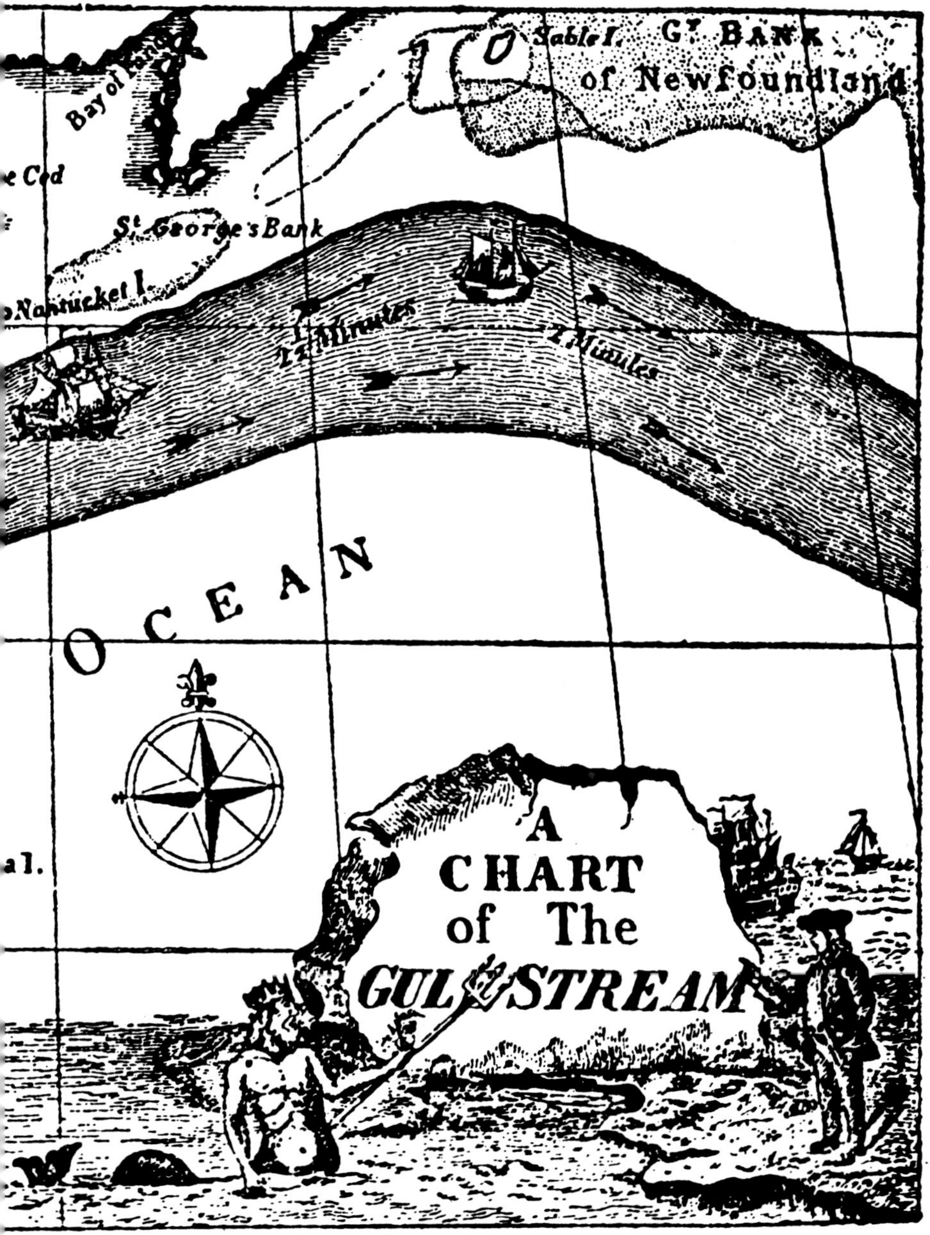
Sable I.
Gt. BANK
of Newfoundland
Bay of Fundy
Cape Cod
St. George's Bank
Nantucket I.
Minutes
Minutes
OCEAN
al.
A
CHART
of The
GULF STREAM

Benjamin Franklin (Mansell Collection)

Hernando Cortes (Mansell Collection)

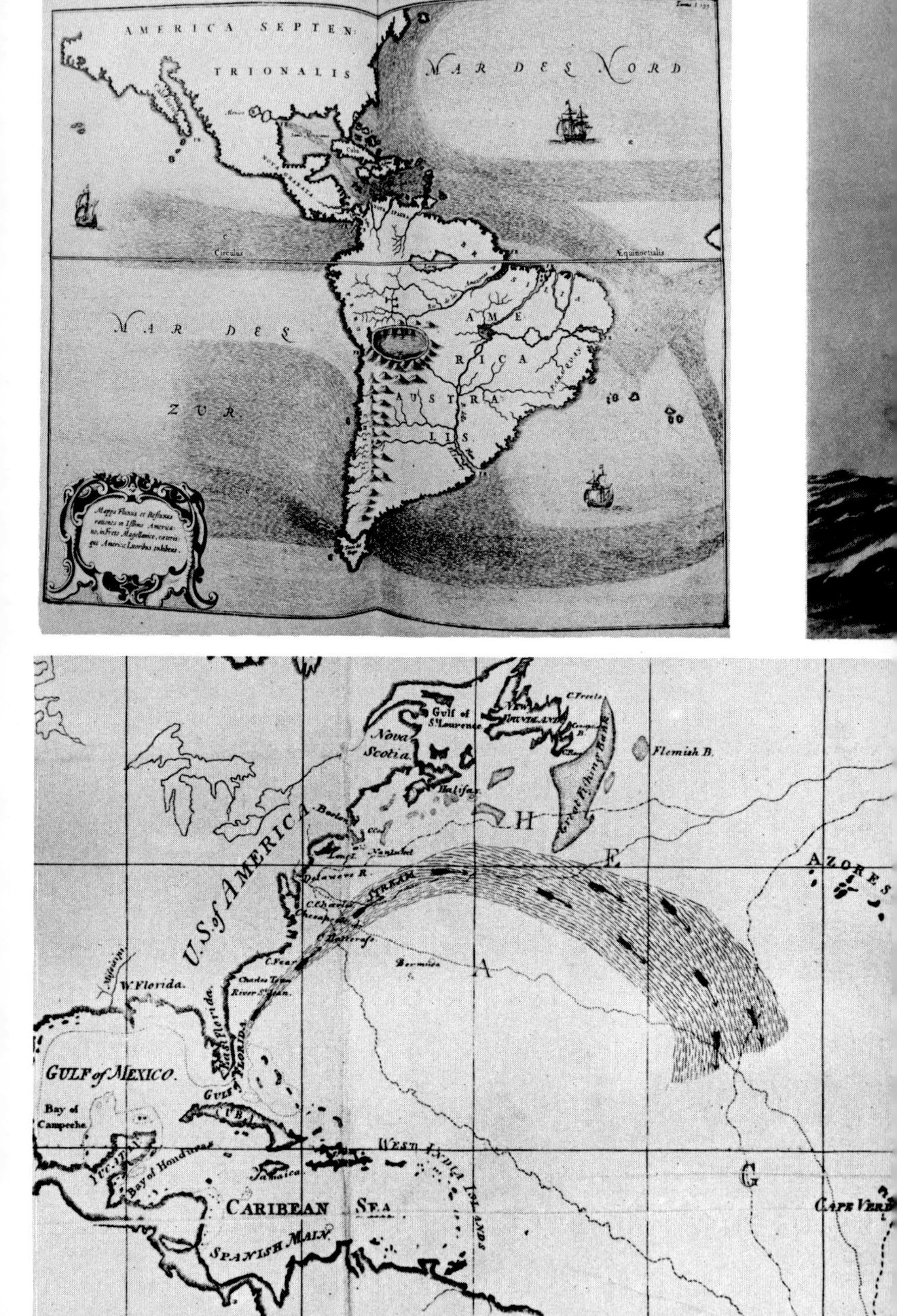

AMERICA SEPTEN: TRIONALIS
MAR DEL NORD
Circulus
Æquinoctialis
MAR DEL ZUR
AME RICA AUSTRA LIS
Gulf of St. Lawrence
Nova Scotia
Halifax
Boston
Flemish B.
Great Fishing Bank
AZORES
U.S. of AMERICA
Delaware R.
STREAM
C. Charles
Chesapeak
C. Hatteras
C. Fear
Bermudas
W. Florida.
Charles Town
River St. John
East Florida
GULF of MEXICO.
Bay of Campeche
CUBA
Jamaica
Bay of Honduras
WEST INDIA ISLANDS
CARIBEAN SEA
SPANISH MAIN
CAPE VERD

Above left: *map of America, showing ocean currents in the Pacific and Atlantic Oceans, including the beginnings of the Gulf Stream. From Athanasius Kircher's* Mundus Subterraneus *of 1678* (Ronan Picture Library)

Below left: *section of the Atlantic Ocean showing the Gulf Stream. From Thomas Truxon's* Remarks, Instructions and Examples Relating to the Latitude and Longitude *of 1794* (Ronan Picture Library)

Above: *H.M.S.* Challenger *which between 1872 and 1876 made a large number of oceanographic observations* (Challenger Society)

Above left: *barracudas, predatory fish which are commonly found in the waters of the Gulf Stream* (Photo Aquatics)

Below left: Gaulolepis longidens, *a deep sea fish which occurs in the regions of the Gulf Stream* (Heather Angel)

Below : *goose barnacles, attached to a piece of wood at the surface of the sea, of the type found in the Gulf Stream* (Heather Angel)

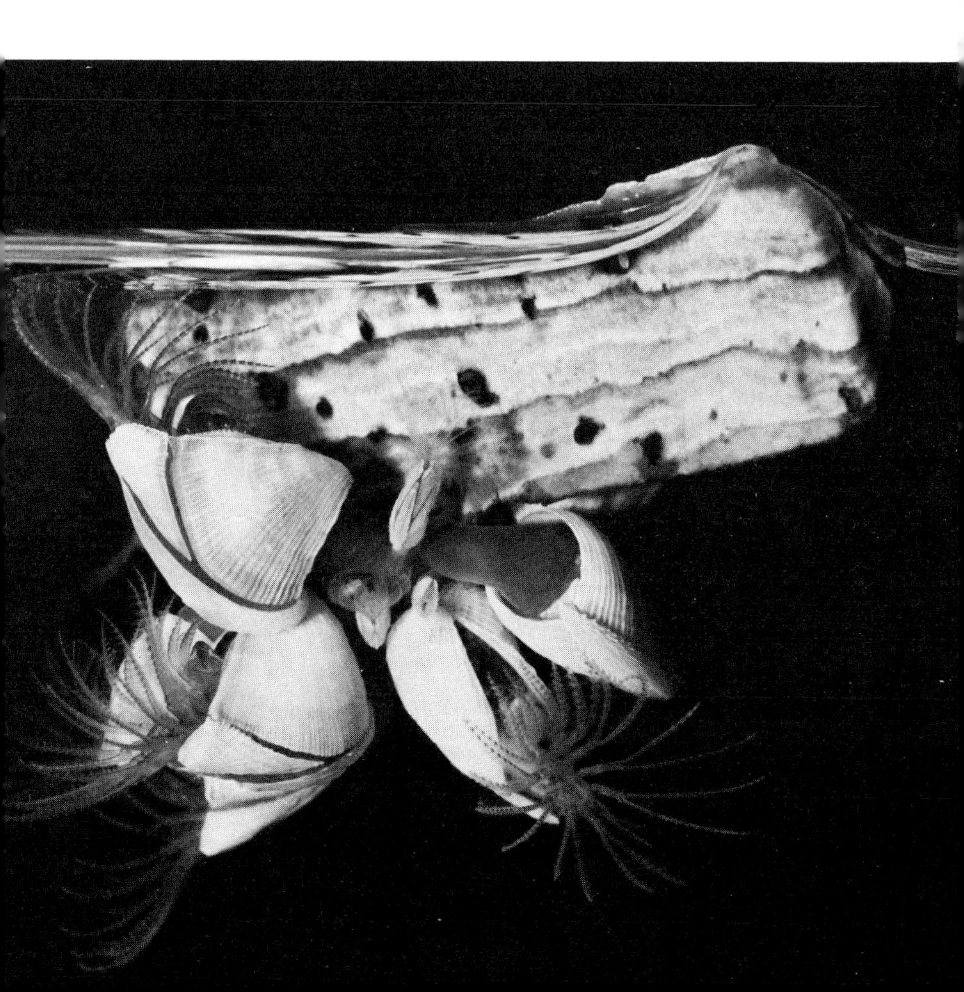

The three stages in the development of the eel. Above: *leptocephalus larvae of eel.* Above right: *young eels or elvers.* Below right: *fully grown eel* (Heather Angel)

The Sargasso Sea from Suhaili (Robin Knox-Johnston)

currents in a west-to-east direction. This current flows at speeds of up to two knots and is caused by the pile-up of water from the two Equatorial currents when they reach the land barrier at the western side of the Pacific basin. The Counter-Current very carefully chooses the doldrum zone of almost no wind which exists between the north and south-east trade winds. This Counter-Current has considerable depth, and is noticed at over 2,000 feet (610 metres) below the sea surface. It therefore carries a great deal of water, and the most modern estimates place it in the size category of the Gulf Stream.

The Pacific differs from the Atlantic in detail, because there is a difference in the land obstacles in the path of the wind-driven Equatorial currents. The North Pacific Equatorial Current has a free run to the west, except for some irritating diversions and hence energy dissipations around groups of Pacific islands and atolls. At Luzon in the Philippines the current is deflected northwards past Formosa towards the coast of Japan. A small amount of the flow turns south along Mindanao to help the build-up of water level that drives the Pacific Equatorial Counter-Current. There is a difference here from the Atlantic régime. The projection of the eastern tip of Brazil is so placed that it deflects part of the southern Atlantic Equatorial current to the north, augmenting the flow of the northern Equatorial current. The Gulf Stream is consequently larger than the Kuro Shio which circulates in the North Pacific ocean. The Kuro Shio hugs the coast of Japan, and at about the latitude of Cape Hatteras in the Atlantic, it turns off from the Japanese coast. It is pushed in this course by the Coriolis forces (see footnote p. 15).

The Kuro Shio is a great current; although its volume of moving water is only about one third the size of our Atlantic Gulf Stream, it is still a warm benevolent current, and like its counterpart, is the outside edge of a general circular water flow in the North Pacific Ocean. The Kuro Shio has a cold visitor from the north, just as the Gulf Stream has its rival Labrador Current. The Oya Shio is a cold stream running southward along the coast of Japan, but it is not so icy as the Labrador Current, and although it produces fogs it does not carry the extra hazards of icebergs. The two currents mix up with the formation of numerous eddies, so that odd patches of warm and cold water can be encountered off the coast of Japan.

The Kuro Shio continues its clockwise circulation by flowing eastwards across the ocean under a new name—the North Pacific Current. There was no great trade between Japan and Asia and the west coast of North America in the eighteenth century, and therefore there was no Benjamin Franklin to follow the Pacific 'Gulf Stream' all the way across the seas, or to collect information on which to draw charts with advice to mariners. The North Pacific Current picks up part of the cold Oya Shio water at Japan so that the northern branch of the west-to-east stream is cold and even goes by another name—the Aleutian Current.

Some comparatively warm water carried by the North Pacific Current turns northward when it impinges on the American Continent, and as the Alaska Current, helps to mollify the climate of the west and south shores of Alaska, in much the same way that the tag end of the Gulf Stream keeps the west coast of Norway cosy in the winter. The Alaskan port of Valdez on the Prince William Sound, a beautiful natural harbour surrounded and protected from the cold winds by high mountains, is ice-free all the year round, and has been selected to be the southern end of the great Trans-Alaska Pipeline, carrying the oil from the Arctic North Slope down to where it is wanted. Valdez is a deep fjord with a narrow entrance, so that it is protected from rough seas as well as from the inclement weather. It has had its troubles in the past, however; in 1964 a large earthquake caused slumping of land, and a subsequent tidal wave. The old town of Valdez was badly hit, but the resolute inhabitants sensibly moved to higher ground across the bay and built a new city.

The clockwise circulation of the North Atlantic is repeated in the Pacific—it is, of course, impossible for the water to do anything else, given the laws of physics and the rotation of the earth. Part of the Aleutian Current sweeps southwards, and skirts the Californian coast as the California Current. This tends to be a cold current compared with what the inhabitants of the sunny west coast of the U.S.A. might expect of their offshore waters. Sometimes north-west winds blow parallel to the coast and cause some upwelling of the colder water through the surface layers that are warmed by the sun. The upwelling tends to bring nutrients to the surface, and there is often an abundance of plant and animal growth. After the summer, the prevailing winds tend to change, and both the upwelling and the cold currents disappear, the California Current moving away from the coast to allow a south-to-north coastal current to develop.

The South Pacific has its circulatory pattern of surface currents, in an anti-clockwise direction. The Southern Equatorial Current, running almost along the Equator forms the northern part of the circulation, while the West Wind Drift or Antarctic Current, which as we have seen has an unobstructed path right round the globe, marks the southern end. The Humboldt Current named after the great scientist, Alexander von Humboldt (see p. 8), is a cold water circulation which flows up the west coast of South America, along the coasts of Chile and Peru. The circuit is completed when the current turns left to join the Equatorial Current. The cold water of the Humboldt Current is experienced on the bathing (or non-bathing!) beaches of Peru because, although the sun is hot enough to warm the surface shallow water, there is an unkind wind which tends to drift this warm water out to sea, and allow the cold Humboldt water to reach the surface. The change from warm tropical waters to the cold Humboldt Current is very obvious to those who take ship south from Panama—soon after crossing the Equator the air temperature drops and there is often an overcast sky—again reminiscent of the effect of Newfoundland in the Atlantic.

There is one small current in the Pacific which sometimes causes disaster off the Peruvian coast. When the sun is south of the Equator a branch of the west-to-east moving Equatorial Counter-Current moves south of the Equator and since it appears at around Christmas-time it is called *El Niño*, or The Child. In some years The Child gets out of hand, and exploring further south gets its warm waters mixed with the cold water flowing up the Peruvian coast. In 1925, for example, the warm *El Niño* reached as far south as latitude 14°S. At first sight this might appear to be a stroke of luck, but the ecological chain of events is such that this excursion of the warm current spells disaster. The cool water is thickly populated with plant and animal life, since it has been fed with nutrients by the upwelling of deeper water. The warm water causes abnormally rapid growth, to such an extent that all available oxygen in the water is used up. The whole area goes anaerobic, just as Lake Erie has done because of excess sewage and industrial waste (see p. 27). Nature, however, in this case does things on a larger scale than human beings have managed to date. Fish die and rot, hydrogen sulphide is given off in the water, the beaches are covered with stinking fish. The white lead paint on a ship's sides is sometimes blackened by the hydrogen sulphide, thus providing the local nickname for the phenomenon as the 'Callao painter'—Callao being one of the coastal towns in the vicinity.

Now that the inhabitants of the earth are at last realizing the effects of modern civilization on the environment, and are waking up to the dangers of fouling the nest irreparably, it will not be surprising to many that this cycle of events does not end at fish. The next in the ecological cycle to suffer are the sea-birds who live on the plentiful fish supplies. When *El Niño* strikes south the birds have a lean season, and either starve, or desert their young in the breeding grounds on the offshore islands. The repercussions even affect man, because a decrease in bird population means a loss of guano which is an important source of fertilizer. As a last straw for the local inhabitants, the warm waters from the north cause violent rains because the air becomes saturated with water to produce tropical storms and heavy rainfall which in turn causes flooding and erosion of the land. Normally this part of the Peruvian coast experiences low rainfall, and the vegetation adapted to this type of climate cannot endure the terrific showers. The only hopeful message that we can extract from this natural conservation disaster is that the system does eventually revert to normal, the fish come back again, the birds follow and the rough winter storms produce waves which clean up the beaches.

A glance at a map of the world, or better still at a globe, will show the difference between the Indian Ocean and the other two larger oceans. The Atlantic and Pacific both stretch from the Arctic Sea to Antarctica, whereas the Indian Ocean is virtually only a southern hemisphere expanse of water, since it is landlocked to the north by the vast continent of Asia. Being only half an ocean has the consequence that there can be only one half of the circulation pattern that exists in the other two oceans. The fact that a large land mass exists to the north will also have an effect on the winds which, as we have seen, form the driving force of the ocean currents.

Below 10°S latitude the Indian Ocean currents follow the normal southern hemisphere anti-clockwise rotation. The South Equatorial Current between 10 and 20°S is a fairly steady wind-driven east to west current which impinges on Madagascar and the African coast. Some of the flow is diverted to the north, but the northern current depends on the seasonal changes in the top part of the Indian Ocean. A large part of the water is deflected to the south to feed the Agulhas Current, which in parts rivals the Gulf Stream in speed. The water carried by the Agulhas Current is more than one third that of the Gulf Stream and speeds up to four knots are encountered. The current runs parallel to the South African east coast and gradually swings round to meet our 'round the world' current, the West Wind Drift. A small part of the Agulhas Current creeps around between the tip of South Africa and the West Wind Drift Current to reach the Atlantic.

The Indian monsoon winds control the currents in the north of the Indian Ocean. During the north-east monsoon, in February and March, there is a normal North Equatorial Current, and an Equatorial Counter-Current flows eastwards between this and the fairly steady South Equatorial Current, in a similar way to the pattern in the middle latitudes of the Pacific or Atlantic Oceans. When the Indian summer monsoon starts to blow the North Equatorial Current is reversed, and is now called the Monsoon Current.

The reversal of water circulation during the year in the Indian Ocean makes it possible to learn more about the effect of the wind on the sea surface. During the International Indian Ocean Expedition, when many nations co-operated to attack all scientific aspects of this previously neglected ocean, one of the most interesting studies was the circulation reversal of the North Equatorial Current. The gathering of a number of ships made it possible to observe the behaviour of the current simultaneously over a large area. The east-west monsoon current in the northern hemisphere summer helps to initiate upwelling of deeper water off the coast of Somaliland, and an area of upwelling was found during the Indian Ocean Expedition to the south of the Arabian peninsula. On the eastern side of the Indian Ocean along the coast of Western Australia and around the maze of islands to the north, the currents are again variable, being controlled by the alternation of trade and monsoon winds.

The circulation of the North Atlantic Ocean has already been described. In the southern hemisphere the circulation follows the same general pattern as that in the South Pacific. Again the circulation is anti-clockwise, but the flow is smaller than in the North Atlantic, because some of the water carried from east to west by the South Equatorial Current was deflected to the north by the north-facing coastline of Brazil to augment the beginnings of the Gulf Stream. The position of the Equatorial Current which is deflected southwards forms the Brazil Current, which eventually swings round to the east to join the West Wind Drift. The Falkland Current, often carrying icebergs from Antarctic waters, squeezes in between the Argentine coast and the Brazil Current to make a nuisance of itself. The circulatory pattern is completed by the Benguela Current which brings cold water northwards along the coast of South Africa. The winds cause some upwelling, which makes the Benguela Current even colder than its southern origin would lead one to expect. The upwelling provides the nutrients which form the basis of a large fish population in the Walvis Bay area.

The pattern of ocean currents cannot have been the same as it is today throughout all geological time. Although the wind system of the world must have existed a great deal longer than the Atlantic Ocean, since it depends on the rotation of the earth, there are causes which may have affected the ocean currents in much more recent times than the millions of years in which we normally discuss the geological past. About ten thousand years ago the most recent of the Ice Ages was covering a large part of the northern hemisphere with an extension of the present Polar ice-cap. The water that was locked up as solid ice piled thousands of feet high on the land had been removed from the sea through evaporation to the atmosphere and subsequent deposition as snow. The sea-level was lowered 300 feet, and this change of level may have had some effect on the course of the surface ocean currents, especially in those places where they run in shallow water. The fingering out of the cold, solid ice may well have deflected the northern part of the Gulf Stream. In its turn, the warm tropical water brought northwards by the stream no doubt fought back to make the ice recede. The details of why Ice Ages occur have not yet been fully explained. We know they have happened because of the evidence they leave behind them, but what upsets the balance to allow the climate to get too markedly out of hand is not properly understood. The circulation of ocean currents certainly plays an important rôle and this matter will be discussed further in Chapter 13.

Some recent measurements relating to the old path of the Gulf Stream have been made by a group who study, as the hydrographers do, the shape and composition of the ocean basin—the bucket that holds the moving waters. The United States Deep Sea Drilling Project, in its 12th Cruise in 1970, bored into the ocean floor at two places north of Newfoundland. Several hundred feet of sediments were penetrated, and sample cores were brought to the surface for examination. At the more southerly station, the mud which had been deposited over millions of years told a very interesting story. The surface layers contained fossils of marine animals and plants that live in cold seas, but deeper down, where the age of the deposits can be calculated to be greater than three million years, there was ample evidence of a warm climate, with its corresponding biological remains. The station farther north showed evidence of a cold climate for millions of years past. The significance of these two borings is that the Gulf Stream, with its warm tropical water, must have been several hundred miles farther north three million years ago than it is today. We do not know whether there was some extra push in the Gulf Stream of these prehistoric days, or whether it was the Ice Ages, first starting three million years ago, which drove the Gulf Stream southwards to its present course.

It is important to remember that we can gain knowledge of the current circulation of the oceans from indirect measurements, such as those concerned with surveying the shape of the sea-bed and with drilling holes through the ocean floor. By adding together all lines of evidence, a full understanding will be gained of the whole mechanism of the important currents such as the Gulf Stream, so that we shall be able to advise on what effect the impinging of man-made alterations to the world will have on natural phenomena. Can we change the Gulf Stream? If so, for good or bad?

The Deep Sea Drilling Project mentioned above is determining the age of rocks beneath the oceans. In the Atlantic one of the objects of the work is to find the time at which North America parted company from Europe, and South America from Africa. Most people today believe that the continents, which occupy one third of the earth's surface, were once all one piece of land, and that they have arrived at their present distribution by a process of drifting, or being shoved apart by forces originating inside the earth. If the continental drift theory is true we can assert categorically that there was no Gulf Stream 150 million years ago. This may sound rather a long time, which it is compared with the normal human life span, but it is a very small fraction of the 4,500 million years of the earth's life as a planet. The Gulf Stream then, although a magnificent river, with the peculiarity that it flows through the sea rather than being confined by earth or rock banks or gorges, is a comparative newcomer on the geological scene.

The currents of the oceans 500 million years ago were probably confined to one large area covering two-thirds of the globe, since according to Wegener's continental drift hypothesis the whole of the land that we see today as separate continents was all in one lump, which Wegener called Gondwanaland. The rotation of the earth will have produced a wind pattern which resembles that of today, and the wind acting on the sea surface will have produced a very long—some 16,000 miles—Equatorial current system, just as it does in the Atlantic and Pacific Oceans today. The Gulf Stream equivalent of those days, or perhaps one should say that the Kuro Shio equivalent, would have been the result of deflection by the land, which in the old Gondwanaland would be what is now the eastern side of Asia, together with the Coriolis forces, and would not probably have had the initial force and concentration provided by the constricted flow of the Gulf Stream through the Florida Straits. There would have been a continuing circulation around the northern hemisphere of the one and only ocean of those days, augmented in the east-west northern section by the equivalent of the southern wind drift of today's Antarctic seas. There would have been an almost symmetrical current pattern in the southern hemisphere, since in those days the earth would not have had a northern hemisphere consisting predominantly of land. However, there would have been a break both in north and south by Gondwanaland itself, so that both the northern and southern drift currents would not have been completely circumpolar as they are in today's southern seas. The return circulation would have been south along the North American coast and north along the west coast of South America, much as occurs in the Pacific at the present time.

There may have been an odd development in ocean currents about 350 million years ago, when Gondwanaland appears to have split into a northern and southern land mass. The southern continent was centred around the South Pole, and there may have been a sufficiently large gap between it and its northern counterpart to allow a complete global circuit for the Equatorial currents. This could have produced an interesting state of affairs, with very rapid current flows when the gap between north and south land masses was narrow. There must, both in the single continent period, and in the early split into north and south, have been great waves produced by the long wind fetches that existed, and the erosion of the coasts by pounding waves must have been extremely severe.

The Atlantic began to form about 150 million years ago, with the east-west split of both northern and southern continents although these splits may have been at different times. The results provided by the Deep Sea Drilling Project and by geophysical researches into the present ocean basins will provide more reliable dates for the geographical happenings of the past. It is sufficient now to state that the Atlantic has gradually widened from a narrow gap to the ocean it is today in a time which is very short compared to the life of the earth. The circulation pattern of the Atlantic Ocean, and the Gulf Stream in particular, are, in geological terms, really newcomers on the world scene.

The Gulf Stream in its present concentrated form at the Straits of Florida must have come into existence at a late stage even in the history of the Atlantic, since firstly a fairly large width of ocean would be needed to allow a strong Equatorial current to build up. It is probable that for a long period there was a northern circulation of the Atlantic which rotated in a clockwise direction much as the Pacific circulation flows today. There would have been a stronger northwards flow on the western side of the Atlantic than the southern flow on the eastern side for reasons associated with the earth's rotation, but there would not have been a four-knot current in the Florida Straits until geological upheavals provided the unique land configuration that allows a pile-up of water in the Gulf of Mexico and a narrow outlet for this water to join the main clockwise circulation of the Atlantic.

The Gulf Stream may well be a transitory phenomenon if we think of the earth hundreds of millions of years hence, because erosion of land or the emergence of new volcanic islands could destroy the Florida Straits and cause the Equatorial currents to follow a new path. There would always be the circulation forced by the wind and demanded by the earth's rotation, but the details dependent on the shape of the ocean boundaries change with the alterations of the land masses moving slowly on the surface of the globe. Modern thought suggests that the Atlantic is still growing wider, although it may have reached a temporary jam because the Americas have run into an opposing crustal movement, but this will in any case not alter the general current circulation as is borne out by the fact that a similar pattern exists in the larger Pacific Ocean. It is much more important for human beings, with their short time scale, to discover the mechanisms of the currents that are in being with the present land distribution, and to make sure that they do not abuse them.

8

STUDYING THE OCEANS

The 1911 *Encyclopaedia Britannica* described the Gulf Stream thus:

> The name properly applied to the stream current which issues from the Gulf of Mexico and flows north-eastward, following the eastern coast of North America, and separated from it by a narrow strip of cold water (the Cold Wall) to a point east of the Grand Banks off Newfoundland. The Gulf Stream is a narrow, deep current, and its velocity is estimated at about 80 miles a day. It is joined by, and often indistinguishable from, a large body of water which comes from outside the West Indies and follows the same course. The term was formerly applied to the drift current which carries the mixed waters of the Gulf Stream and the Labrador current eastwards across the Atlantic. This is now usually known as the 'Gulf Stream Drift', although the name is not altogether appropriate.

What has modern measurement and theory taught us since then?

Franklin's great-grandson, Professor A. D. Bache, was one of the pioneers of modern measurements of the Gulf Stream. Perhaps his interest was aroused in childhood days hearing of his famous forebear and his North Atlantic charts. What better and more interesting addition to the knowledge of the east-west portion of the Gulf Stream which crosses the Atlantic than to investigate the much more rapidly moving source of this current, where the jet of water pours out of the narrows between Florida and the shoals around Bimini? Bache made fourteen sections across the Gulf Stream, locating the position of the current by measuring the temperature of the water. He was the first to notice the existence of the cold streaks in the main flow, which we now know are due to the eddies and meanders which spin off along the course of the current.

Professor Bache was Superintendent of the U.S. Coast Survey (a permanent survey—now known as the U.S. Coast and Geodetic Survey) from 1843 and his instructions to Lieutenant C. H. Davies, U.S.N., captain of the brig *Washington* demonstrate his understanding of the Gulf Stream, and what was required in the way of fresh observations:

> The following questions should be examined:
>
> First. What are the limits of the Gulf Stream on this part of the coast of the United States, at the surface and below the surface?
>
> Second. Are they constant or variable, do they change with the season, with the prevalent and different winds; what is the effect of greater or less quantities of ice in the vicinity?
>
> Third. How may they best be recognized, by the temperature at the surface or below the surface, by soundings, by the character of the bottom, by peculiar forms of vegetable or animal life, by meteorology, by the saltness of the water?
>
> Fourth. What are the directions and velocities of the currents in this Stream and adjacent to it at the surface, below the surface, and to what variations are they subject? What peculiar arrangement of the currents takes place at the edge of the Stream in passing from the general waters of the ocean into those of the Gulf? Some of these questions will require long-continued observations to solve. If you can obtain something like approximation to the normal condition of the Stream in this summer's work it will be quite satisfactory. Make, then, as many cross sections of the Stream as convenient and as the investigation may show to be necessary. In these sections (1) determine the temperature at the surface and at different depths; (2) the depth of water; (3) the character of the bottom; (4) the direction and velocity of the currents at the surface and at different depths; (5) as far as practicable notice the forms of vegetable and animal life.

Professor Bache suspected that there might be an opposing cold current under the surface stream and his instructions continue:

> The existence of a counter current of cold water from the

poles below the warm current from the equator has been supposed. This current would produce a position of rest, in which if a heavy body attached to a light one at the surface were immersed, the light one would drift off down the stream of the superior current. If a light body were sent down to the counter current and then detached, it would rise at a point up the stream of the surface current. A boat might be anchored on it by attaching to it a body which would produce a considerable resistance to motion. Two boards put together crosswise would answer the purpose well. It may be that if there is no counter current the velocity near the bottom is so much checked as to cause a variation to be discernible in some such way.

The latitudes of Florida are well known to be subject to hurricanes, and in 1846 the *Washington*, now under the command of Lieutenant George M. Bache, U.S.N., was nearly wrecked. The captain and ten of the crew were swept overboard and lost, and the vessel barely managed to reach port with that season's observations. Lieutenant Bache gave the name 'cold wall' to the rapid change in temperature which is found in many parts of the western side of the Gulf Stream.

The U.S. Coast Survey continued to investigate the Gulf Stream. The 1877 annual report comments on the work of the *Blake* with Lieutenant-Commander Sigsbee as captain and Lieutenant Pillsbury as his executive officer.

The log of the *Blake* shows that the vessel rode out more than twenty gales while working in the Gulf. The results include soundings, temperatures, specimens of water from the surface of the sea and at intervals downward, specimens of the bottom and records of the directions and force of the currents, making an aggregate unequalled in the annals of ocean physics by results obtained in the same limit of time, by any party employed by any nation.

The American survey officers and crew appear to have been somewhat more energetic than the complement of H.M.S. *Challenger* which had made observations a few years previously in the Gulf Stream area, while on the great 1872–6 round the world voyage which set the pattern for world-wide oceanographic study.

> Without specifying the great results obtained from the continuous research, I may be pardoned in referring with some gratification to the fact that in the small steamer *Blake*, of only 320 tons burden, new measurement, under the energetic and skillfull command of Lt. Cdr. Sigsbee and Commander Bartlett, with a complement of forty-five including officers and crew, five soundings were taken in the depth of 2000 fathoms during the same length of time which the *Challenger*, of 2300 ton burden, with a complement of 29 naval and civilian officers and a correspondingly large crew, occupied in taking one, and 5 hauls of the dredge at the same depth were made by the Blake in the time occupied by the Challenger in making one.*

A little friendly rivalry and competition are still useful in speeding up progress of these difficult observations at sea.

The classic feat of these times was performed in 1888 and 1889 by Lieutenant J. E. Pillsbury, who succeeded in anchoring the U.S. Coast and Geodetic Survey vessel *Blake* for days on end in the Florida Straits where the Gulf Stream, with a four-knot flow, is at its strongest. Instead of observing the water flow by the drift of the ship it was possible to obtain a series of direct velocity measurements at different depths. During six months in 1889 the *Blake* steamed 11,850 miles, making 2,577 current measurements and 2,535 temperature recordings. The ship was anchored 39 times in various depths of water up to 13,780 feet (4,200 metres). He lowered current meters to various depths; these meters were held vertical by means of a horizontal attachment to the anchor cable of the ship. The anchor cable needed was two to three times the water depth on account of the strong currents. In all, Pillsbury made observations of five sections across the Florida Straits and at a station off Cape Hatteras to the north, and he also studied the relative strength of the various streams in the passages between the Windward Islands.

The Pillsbury measurements of water velocity and temperatures right across the confined stream in the Florida Straits have provided one of the best checks of the theory by which current patterns are calculated from the water pressure at

*From annual report of U.S. Coast & Geodetic Survey, 1879.

depth. The most effective way of deducing the circulation of water in the oceans is to measure temperature and salinity at as many places as possible, and from these calculate both the surface and deep water currents (see Chapter 7). Normally the definition of the edges of the moving masses of water is not very exact and so the calculations of actual speed of flow are somewhat vague. In the Florida Straits, however, the Gulf Stream is contained by a known, surveyed profile of sea-bed, just as a river on land is confined by its stream-bed. The measurements made by Pillsbury embraced the whole cross-section of the Straits both sideways and down to the bottom, so that the total flow of water through the Straits could be calculated.

Dr. Georg Wüst, the famous German oceanographer who made extensive oceanographic measurements in the research ship *Meteor* in the 1920s, used Pillsbury's measurements in a direct comparison with the deductions from salinity and temperature observations in profiles made by the *Meteor* across the Gulf Stream. The Coast and Geodetic Survey had prepared four temperature and salinity sections across the Florida and Antilles streams with the survey ship *Bache*, while the Norwegian ship *Michael Sars* had made a modern hydrographic cross-sectional profile south of the Newfoundland Bank in 1910. The Danish *Dana* expedition under Johan Schmidt in 1921–2 provided further transverse profiles of the Gulf Stream, and together with the *Meteor* and the earlier results allowed Dr. Wüst to make estimates of the discharge of water in the Florida Straits. The following, in a rough translation from Georg Wüst's inaugural lecture at the University of Berlin on 15 November 1929, gives the results of the combination of theory and direct current measurements. These confirm the picture that has already been given of the enormous flow of water in the Gulf Stream, and the rather exceptional jet action caused by the restricted Florida Straits channel in that part of its course which is very much like a river.

The Vertical Velocity Distribution

A complete idea of the dynamics of the Gulf Stream is provided by the three velocity profiles through the Florida Straits based on the direct current measurements of Pillsbury and the profile which has been derived by me indirectly by dynamic calculation from the temperature and salt content distribution. Initially we get the impression of a mighty river profile. The distribution of the velocities in the uppermost 200 m. and their absolute magnitude are similar to the river profiles that we know from the large continental streams on land. However, the physical conditions and dimensions are completely different. The river flows as a gradient stream in a bed inclined and that generally enlarges. The bed of the Florida Straits increases in the direction of the stream and is constricted from the entrance of the straits continuously almost up to its opening. It is not, of course, fully filled by the current everywhere. However, the temperature and salt content sections indicate that the lower limit of the stream, too, increases in accordance with the bottom profile, even if to a lesser extent. The current direction of a river is shown in increased degree by the curvature of the bed. The axis of the Florida stream, corresponding to its formation as a pressure stream, selects the shortest path. Jet-like, as in the Bosphorus, the stream shoots through the narrows of the Straits and retains the direction forced upon it for a long distance beyond the narrows. There is another difference: in rivers, which after all contain homogeneous water masses, the distribution of the velocity is governed mainly by migrating vortices which are formed at the bottom and rise to the surface. In the case of the Florida Stream, on the other hand, we encounter strong stratification at depths below 200 m. This stratification greatly inhibits turbulence. This explains the close relationship between stratification and velocity distribution. For we have seen in our sections that the isotachs in the lower layers (below 200 m.) are oblique in the same sense as the isotherms and isohalines.

Characteristic for the force of the pressure forces that generate the Florida Stream is the fact that even in the first part of the Sea Straits, which are still wide open, velocities of more than 120 cm/sec are achieved on the surface and of more than 80 cm/sec are achieved at a depth of 200 to 300 m. The compression of the flow lines in profile III in the Narrows of Bimini means that extraordinarily high mean current strengths of approximately 180 cm/sec occur. In individual cases, the velocities frequently increase to 200 to 250 cm/sec values equalled only by the big continental rivers in flood. A clearer idea of these tremendous current strengths is obtained if it is remembered that in the Straits of Bimini a ship would be moved forward by 200 km. in 24 hours without any effort of its own. While in most ocean currents, especially in the wind drifts, as for example in the mighty equatorial currents, the velocities rapidly reduce with depth, opposite currents being encountered at a depth of approximately 100 m., the water masses in the Gulf Stream, in conformity with its formation as a pressure stream, are transported down to depths of 400 or even 600 m. at relatively high speeds in the direction of the surface current. In this respect, the Gulf Stream occupies a special position among the sea currents. At the narrowest point of the Florida Straits, the water masses in a cross section approximately 70 km. wide and 200 m. deep are moved forward at a speed of more than 1 m/sec. The tremendous dimensions of this flow are realised only if it is noted that the length of the Florida Straits corresponds approximately to the length of the Rhine from Lake Constance to its mouth while its width is equal to the width of the lands of Baden and Alsace together.

The success of Dr. Wüst's comparisons between direct measurements of speed of flow and the currents calculated from temperature and salinity measurements have led to an extension of our knowledge of the Gulf Stream by many more observations of the water-bottling type (see Chapter 7). Generally, where direct current speeds are measured, the deductions from the pressure picture of the oceans are confirmed. Some of the latest techniques for studying ocean circulation rapidly over large areas are confined to observations of the water surface, but enough is now known of the laws that govern the whole phenomenon to enable translation into meaningful results for the whole body of water.

The early observations of the dead reckoning type were supplemented in the last century by many experiments using drifting floats. In the late nineteenth century, Prince Albert of Monaco, who spent a great deal of his time and of his personal wealth in encouraging the new subject of oceanography, was one of the foremost contributors both to the design of floats and to the results obtained with them. The voyages of the Prince's ship the *Hirondelle* during the four years 1885–8 were concerned with the surface currents of the North Atlantic in addition to the work on the shape of the ocean floor and the animal and plant life in the sea. The development of a suitable float that would faithfully follow the surface water movement and not be taken in false directions by local winds, can be traced today at the Monaco Oceanographic Museum, where Prince Albert's nets, fish traps, bottom samplers and other original and ingenious oceanographic apparatus are preserved for all to see. The float that was finally adopted was a glass bulb sheathed in a copper envelope and ballasted with a suitable quantity of lead shot to enable it to float with virtually no surface for the wind to affect. The floats contained a message written in nine languages requesting the finder to inform the French Government of the date and location of retrieval of the float. By studying the pick-up places of floats launched in various parts of the sea, it was possible to deduce where the main streams in the sea surface were moving, and an estimate of average speed of the currents could be obtained from the known time of travel.

In making surface current observations from a ship at anchor, various types of float such as corks on a light line, or partially submerged poles to give an average current over the top ten feet of water, are paid out from the ship and the time measured for a certain distance of travel. Currents at depth are more difficult to measure, and many devices have been developed over the past fifty years or more. As with most oceanographical equipment, results depend to a large extent on how well the apparatus is maintained and used. Unfortunately, it is still not uncommon for watertight instruments to return to the surface full of water, or to have the supporting cable wrapped round the propeller. An expensive underwater camera was once lost in the calm waters of Gibraltar harbour while many trailed magnetometers have disappeared, possibly in this latter case as a tempting morsel to a shark.

Dr. J. N. Carruthers, who has been measuring currents and

other properties of the oceans for many years, has devised a simple but adequate current measuring device which is so easy to use that it is virtually foolproof. The mechanism, if it can be called such, consists of a bottle partially filled with a jelly. The jelly is warmed up so that it melts before lowering the bottle to the sea-bed. The bottle floats above the sea-floor, held stationary by a sinker, and is pulled at an angle by any current that is present. The cold water makes the jelly set solid after a while, and when the bottle is recovered, the solidified jelly provides a measure of the tilt of the bottle when it was pulled out of the vertical by the current. A small compass needle floating in the jelly is used to 'freeze' the direction in which the bottle is pulled by the current, and so gives the bearing of the current. The jelly bottles have proved to be of great help in offshore oil production, where rapid surveys of the currents at the sea bed are needed for planning underwater pipelines, and for controlling diving conditions around offshore drilling rigs.

The simple, old-fashioned current measuring methods still provide a great many useful results. In the 1950s interest was aroused, mainly by the disgorging from old wartime wrecks of oil which sometimes ended up as a horrible mess on holiday beaches. Oil was also known to be discharged at sea when cleaning out fuel tanks of passenger and cargo ships, or when flushing out the ballast tanks of oil tankers. The ocean is very large, some 300 million times the volume of oil produced annually, but since oil floats, and the ocean surface is moving, it is possible for some oil to reach the shore before it becomes completely lost by mixing with the vast volume of sea-water, or being broken down by bacteria in the sea. Some examples of lumps of heavy oil have been recovered from the sea in which the oil was preserved as a heavy black mess by a solid coating of oxidized or weathered oil. This outer coating in some cases had barnacles and other small marine parasites living on it. While at sea these floating homes cause no trouble, but when they are thrown up on the beach by waves they break to produce the thick black mess so hated by mothers of small children.

In order to find out whether there was a safe distance from land at which oil waste could be dumped in the oceans, a large-scale experiment was arranged in the north-east Atlantic by Dr. George Deacon, Director of the (British) National Institute of Oceanography. A large number of surface floats in the form of plastic envelopes were dropped hundreds of miles out at sea, and their final destinations and time of arrival were determined when they were returned to the Institute of Oceanography in order to collect the reward promised on the label of the float. This old method of current observation showed the general pattern of flow, to some extent part of the Gulf Stream circulation, and demonstrated that no oil should be dumped within 200 miles of land. This injunction became part of the 1958 Convention of the Sea, and today even stricter self-imposed regulations have been created by the oil industry.

The world production of crude oil in 1970 was around 2,000 million tons, about one cubic mile in volume, and about one-fifth of this oil was carried up the English Channel. Twenty years ago, with production less than one quarter of today's figure, the sea transport of oil produced no pollution problem. Now even small spillages of one per cent mean quite a large tonnage of oil out of the vast quantity that is shipped. An oil tanker is usually a one-way carrier in that when it has delivered its cargo it has to travel back to the oilfields empty, or rather in ballast, to collect a fresh cargo. In order that the tanker shall not be too far out of the water, and therefore unstable, on the return journey, sea-water is pumped into some of the tanks. This sea-water has to be discharged before the tanks can be refilled with oil, but the water cannot be pumped out at the loading terminal because it would cover the harbour with a few hundred tons of heavy oily sludge which had adhered to the sides and bottom of the tank, and had been washed off by the water ballast. In the old days when the traffic was small, the practice was to wash the tanks when the ship was over 200 miles out in the deep ocean, and dump the oil washings overboard. The National Institute of Oceanography's drift experiments showed that the Gulf Stream and its 'tail end' currents could bring some of this oily mess back to shore, and therefore about ten years ago the oil companies decided that a new washing technique would have to be devised. This is how 'load on top' originated.

Instead of pumping the tank-washing water, contaminated by oil sludge, into the sea, all washings are put in the end tank. After a few days most of the oil floats to the top of the water, and clean water beneath can be pumped into the ocean. The mixture of sea-water and oil that remains is left in the end tank and when the ship is loaded, the new cargo of crude oil is loaded on top of this mixture. The only people to suffer are the refineries, since on discharging the cargo some salt water enters the refinery with the crude oil. Modern refineries are equipped with desalinating plants, vital to stopping corrosion in the refinery, and although these plants add to the cost of oil-refining it is a worthwhile price to pay for keeping the sea surface free from oil. The only oil that spills in the sea is from the washing out of fuel tanks of cargo ships, or the occasional accident, when a tanker is in collision or runs aground. New cargo ships are being constructed with oil/water separators, and large ports are providing facilities for accepting oily wastes, so that, in spite of the enormous increase in consumption of oil, the 'oil on the sea' problem has been brought under control. Accidents do occur at sea, but research into containing oil spills and dispersing the oil at sea and on beaches has been carried out for many years, and methods are available for dealing rapidly with any incident.

The movement by currents at the sea surface is often confused by the effects of wind, and circulation of the ocean surface water is better observed at depth. An interesting freak case occurred in 1892 when the schooner *Fred Taylor* was cut in half by the steamship *Trave* about 100 miles to the south-east of Nantucket Island. The two halves of the schooner floated for about six months, but in quite different directions; the front part went northwards, and finally grounded on the coast of Maine, whereas the stern travelled south-west and sank in the mouth of Delaware Bay, some 435 miles from the bow section. Apparently the stern section settled much more deeply in the water than the forward half, and was carried mainly by currents rather than drifting under influence of the wind.

Although sea surface temperatures are very useful in following the movement of warm currents like the Gulf Stream, the stirring up of the sea by waves makes measurements below the surface, such as taken with the bathythermograph (Chapter 7), of greater value in research. A new continuously recording instrument for measuring the temperature of the top 1,000 feet or so of the oceans has been in use for the last ten years. A long chain of electrically recording thermometers, or thermistors, is trailed beneath the oceanographic ship, and marks on a recorder on board ship the curve of temperature against depth as the ship traverses the currents. This was one of the methods used in a multi-ship survey of the Gulf Stream carried out by Woods Hole Oceanographic Institution in 1960. Dr. F. C. Fuglister has written an account of this work which covered an area of about half a million square miles, extending from the continental shelf south to the latitude of Bermuda, and from the Grand Banks of Newfoundland west to Georges Bank, off Cape Cod.

In the first phase of the 1960 experiment a grid of stations, where temperature and salinity profiles were measured, was arranged to cover the entire area, to determine the pressure map in the water. The second phase was to make direct current observations in the Gulf Stream, and below the stream. The way in which the latter measurements were made was most ingenious, and was the invention of Dr. J. C. Swallow of the British National Institute of Oceanography.

The 'Swallow' float is constructed of aluminium alloy tubes, and is so weighted that it will sink to a certain pre-determined depth, and will remain drifting with any currents that happen to exist at this depth. The aluminium tubes sink because they are weighted to be heavier than water at the surface. As the float sinks deeper it is compressed by the pressure of the water around it, but because the tube is less compressible than the sea-water that surrounds it, there comes a depth at which the water becomes just dense enough to support the tube. The float is equipped with a 'pinger' which transmits an acoustic signal to a ship which listens with hydrophones. The ship can thus tell the direction and bearing of the float relative to its own position. In order to avoid the need to anchor the ship to provide a fixed reference point, the ship periodically positions itself by radar relative to a buoy which is anchored to the sea-bed by a wire. The buoy swings on the end of the three-mile wire as surface currents and winds vary, so that yet a third fixed reference is used. This is the sea-bed itself. Every few hours the buoy position is checked by making careful soundings of the sea-floor all around it. As the buoy drifts, the whole pattern of the sea-bed will appear to have shifted position.

Since the sea-bed is known to be fixed, the movement of the buoy can be determined, and corrections applied to the apparent movement of the Swallow float relative to the ship.

Surface measurements of current can now be made from a ship under way, as well as by the conventional float methods from an anchored ship. The geomagnetic electro-kinetograph, or GEK, has two electrodes which are towed through the water. A voltage develops between these electrodes if they are in a moving conductor which is itself in a magnetic field. The current of sea-water provides the moving conductor, and the earth supplies the magnetic field, so that the electrode voltage can be converted to speed of movement of the current under investigation. It is necessary to trail the electrodes in two separate courses at right angles in order to calculate the direction and speed of the current.

The 1960 experiment confirmed the general principles of flow of the Gulf Stream, but the observations, as is so often the case, left many more questions to be solved in the future. It had been known from previous experiments in the Bermuda area, that a southward current existed below the main northward flow of the Gulf Stream. The early work of Wüst and Stommel had forecast such a reverse current to account for the temperature and salinity measurements and the water transport that can be calculated from them. In 1957 confirmation of the southward current was obtained by direct observation with Swallow floats at 9,190 feet (2,800 metres) below the sea surface. The speed of the current was only a fraction of a knot, but this is sufficient to move large quantities of water about the world. It is probable that this deep counter-current to the Gulf Stream is not always directly beneath the fast-moving surface current. To the north in latitude 38°N, deep Swallow floats showed a north-east current at depths of 6,600–9,850 feet (2,000–3,000 metres), and this may be interpreted as a complicated deep-water circulation; but in the area east of Cape Hatteras, the deep water is moving in the same direction as the surface Gulf Stream.

The continuing work that takes place on the Gulf Stream indicates that the detailed behaviour is still not fully understood. The picture is confused by the meanders that take place, but some order is being established even with these random movements as more regular observations are made. The meanders are certainly of much smaller amplitude to the south of 65°W, that is, the longitude of Bermuda, than they are farther downstream of the current. One series of modern measurements which may throw more light on this aspect of the water circulation are those using infra-red to give instantaneous pictures of the position of the Gulf Stream from aircraft and satellites. The use of infra-red was first tested by Stommel and Parson in 1953, from an airplane flying at 1,000 feet above the sea. They found contrasts of temperature when flying across the Gulf Stream of 18°C. (32°F.) in late summer and 11°C. (20°F.) in late winter. The U.S. Naval Oceanographic Office later developed an airborne radiation thermometer to operate 500 feet above sea surface to measure sea surface temperatures to an accuracy of ½°F. Regular flights over the path of the Gulf Stream allow changes in course of the current to be observed. For example, in the month of February 1966 the eastern part of the Stream, where it goes to the east across the Atlantic, moved sixty miles to the south, although the course of the west and south part of the Stream was steady at this time. The airborne observations show clearly how the meandering of the Gulf Stream departs from the mean normal path of the Stream, and helps to account for the wide swathe of Gulf Stream marked on Benjamin Franklin's charts.

The average of 3,600 aircraft observations of the northern edge of the Gulf Stream have been compiled to produce a chart showing the upper and lower limits of the current's north wall. Although accurate surface temperature measurements from infra-red air observations are only possible if ships are present to give on-the-spot calibrations, as is arranged in large-scale oceanographic experiments, the contrast in temperature at the northern side of the Gulf Stream is, as we have seen in earlier chapters, so large that it shows up as a relative change in temperature even if the absolute values are not known. The Gulf Stream certainly wanders about during the course of a year. In general, the sharp northern edge of the current will meander about the mean course for the year by more than 100 miles north or south. Whether this is the effect of something further upstream, such as variation in the flow through the Florida Straits, or whether it is the result of cold water and floating ice from the north, is not yet known. We do not know whether the meanders of the Gulf Stream affect the local weather in Europe, but we are learning gradually by the use of better equipment and by more comprehensive experiments.

One of the best modern methods of learning more about the weather has been photographing cloud patterns from satellites; in fact this work will in the future almost certainly be regarded as of far greater importance than landing on the moon. It is not surprising, since the meteorological spacecraft are now well established, that they have been enlisted to supplement the infra-red studies of the Gulf Stream. High resolution infra-red pictures were recorded, for example, in 1966 by the Nimbus II satellite. The cameras recorded a swathe 2,180 miles (3,500 kilometres) wide round the earth, and were accurate enough to give 7 miles (11 kilometres) resolution. This was adequate to show the large meanders of the stream, and these were in good agreement with aircraft and surface ship observations. The only snag of this method of watching the Gulf Stream is that it is a fair-weather operation in that cloud cover obscures the sea surface sometimes. On the other hand, satellites orbit for long periods of time, and in the course of years of observation should provide new information to supplement that obtained from ships and aircraft, which also have their own limitations in respect of complete coverage of the Gulf Stream pattern.

The flow of water through the Florida Straits has been monitored by an electrical method similar to that used in measuring currents from a ship under way by the GEK towed electrode technique. The Western Union telegraph cable connecting Key West in Florida to Cojimar in Cuba provided a suitable link to join specially matched GEK electrodes placed at either end. The flow of conducting sea-water in the earth's magnetic field produced a voltage difference between the two ends of the cable, and was recorded for the year 1952–3. The average quantity of water which streamed through the Straits was found to be 27 million cubic metres per second, which is in good agreement with that deduced from Pillsbury's direct measurement of current (28 million cu. m./sec.) and also with the figure of 26–30 million cu. m./sec. obtained from consideration of the flow pattern obtained from temperature and salinity measurements.

The continuous recording of flow allowed some information to be gained regarding the variations of flow from one month to another. In December 1952 there was a doubling of flow in a period of only three weeks, while the overall variation in the year was from a minimum of 14 million cu. m./sec. to 39 million cu. m./sec. In general the flow was low (around 20 million cu. m./sec.) for the months August to December, and high (about 30 million cu. m./sec.) from January to May. The flow rate would be expected to depend on the total effect of the drive exerted by the wind across the Atlantic, and good average agreement between wind over the Atlantic and flow has been obtained by Professor Munk. The electrical measurements did not show any direct correlation between Atlantic wind strength and monthly variations in flow through the Straits; this is not surprising, because of the complexity of the wind pattern over a large stretch of water such as the Atlantic. The difficulty of adding up all the individual wind contributions is made almost impossible at present by the paucity of continuous observations *en route*. A more interesting correlation indicates that the wind off the coast of Florida and the south-eastern United States controls the flow rate, presumably by pulling the water through the Florida Straits instead of having it pushed by the Atlantic current. The flow would be expected to be affected equally by a pile-up of water on the entry side of the Straits as by lowering of level at the exit. This is demonstrated by the agreement between short period variations of flow and the tidal pattern observed on the Atlantic side in Miami and the Gulf of Mexico side at Galveston.

As new facts concerning the Gulf Stream flow are gathered using the new methods of deep floats, air observation, and more temperature, salinity and current measurements from ships, the theory of water movement in the oceans can be built up into a more comprehensible picture. One of the facets of the system that has been considered is the effect of the shape of the sea-bed on the Gulf Stream floor. In the early part of its course the Stream flows over the Blake Plateau and into deeper water. However, east of Cape Hatteras the 1960 experiments showed deep water flowing in the same direction as the main surface current. This must have come from some source other than the southern part of the Stream, and perhaps is due to deep water from the east in latitude 33°N to 36°N which turns north and then east to follow the main Gulf Stream flow. This deep water flow is deflected at about 39°N from an easterly to a northerly direction by the large obstacle provided by Kelvin Sea Mount. It has been suggested that the pattern of the meanders in the area east of Cape Hatteras may be initiated by the Stream bouncing off the continental shelf to give a series of eddies as it flows up and down the continental slope.

Some new thoughts have been produced concerning the part that meandering of the Stream plays in dissipating energy. By analogy with the atmosphere, where eddies may sometimes have the reverse effect and generate energy, it appears possible that meanders may provide a previously unexplored mechanism for driving part of the Gulf Stream. These hypotheses will lead to the carrying out of specific experiments to check the validity of the predictions, and new methods of measurement will in turn be invented to cope with the new problems posed. Useful results over long periods of time may well be obtained in the future by the continuous operation of current meters suspended from anchored buoys.

One interesting question which must occur to anyone who looks at a world map of surface currents is why the currents, such as the Gulf Stream in the Atlantic and the Kuro Shio in the Pacific, are so much more powerful on the west than on the east side of these oceans. A lack of symmetry in the ocean circulation might be expected because the earth rotates one way and not the other, and undoubtedly the effect of the wind stream in low latitudes being predominantly from east to west must in some way account for the strong west-side currents. However, there is more to it than the mere wind effect, because the play of forces on the western side is different from that on the eastern, on account of the Coriolis forces, which are themselves a consequence of the earth's rotation.

It is opportune for those who study the Gulf Stream in particular and ocean circulation in general, that an International Decade of Oceanography has been started. As will have been gathered from the descriptions of Gulf Stream processes in previous chapters, we do not yet understand what is going on in the three miles of water that cross a large part of our earth. The effect of eddies and meanders in the Gulf Stream may be much greater than has been allowed in the classical theory of currents, and it may be necessary to consider these phenomena as part of the main process rather than as side-effects. This change in thought has already occurred in investigations of the circulation of the upper atmosphere, where rather than playing a dissipative role, the large-scale fluctuations have been shown to be instrumental in driving the main circulation. We seem to be at a chicken and egg philosophy, which may also apply to the part the Gulf Stream plays in changing the climate, or vice versa.

The flow of deep water, and the time for changing surface water to deep and back again, which as we have seen affects our behaviour with regard to waste disposal is bound up with the theories of water movement that are now being reviewed. The Gulf Stream, because it is a much studied phenomenon, and because it is a conspicuous current—and incidentally near the U.S. eastern seaboard where many famous oceanographic schools have their headquarters—may well prove to be the main experimental ground for much of the new work of the Decade of Oceanography.

There are proposals for a detailed experiment aimed at finding out the laws which govern the exchange of energy between sea and air, and the effect this has on the energy pattern of the water beneath the surface. The Scientific Committee for Oceanic Research, an international body which brings together scientists of all disciplines who study the oceans, has set up a working group with the self-descriptive title, Mid-Ocean Dynamics Experiment, or MODE. The aim of the working group is to plan experiments for the 1970s and 1980s in order to provide observational data about medium-scale eddy processes in the ocean that can be related realistically to a mathematical picture of the circulation of the ocean. The results of this research will be the next step towards a better understanding of the ocean currents, and hence of the climate, weather, spread of pollution and marine biological activity.

The MODE project is the brainchild of Professor Henry Stommel, who has already made great advances in the field of ocean water circulation. It is probable that several smaller preliminary investigations (PRE-MODES) will be needed before the large experiment is planned, and PRE-MODE I is scheduled as a pilot study to take place in late 1972, probably in a 120-mile square south of Bermuda. The main MODE will probably encompass the whole North Atlantic circulation, and should really tidy up the theory of the Gulf Stream behaviour. The MODE series of experiments will call for new types of measurement, and improvement of existing apparatus. Moored current meters and bottom-mounted current meters will be used in large numbers, together with sensitive bottom-mounted pressure recorders. Neutrally buoyant floats will be tracked by means of underwater acoustic signals, and oceanographic ships will make routine temperature and salinity recordings in the selected area.

It is important that this type of experiment should be supported by as many governments as possible, because of the large total effort required, especially in the main series of observations which should take place in the late 1970s when the best apparatus and measuring techniques have been established by the preliminary experiments. It is much more important for the human race to spend large sums on acquiring knowledge which leads to a proper understanding of how the world works than it is to mount spectacular events such as landing humans on the moon. It is hard to convince governments of this truth, but with the world tending to be spoiled by the rapid increase of human beings and their voracious demands, it is time we attacked the rather dull and more mundane problems of science and stopped playing to the gallery of television and the popular press. The circulation of the oceans may well hold the key to our existence in the future, and projects like MODE should be encouraged to the full.

9

SERPENTS AND SUCH

The hub of the great circulating wheel of water in the North Atlantic, of which the fast moving Florida Current and the Gulf Stream form the containing rim to the west, is a warm, deep volume of slowly moving water known as the Sargasso Sea. The warm surface water is at its thickest in this central area of the North Atlantic circulation and the Sargasso Sea forms a stable, slow-moving focus around which the Equatorial Current, the North Atlantic Drift, the Gulf Stream, and the Canaries Current can rotate. In the fifteenth and sixteenth centuries, when the Atlantic was being discovered by sailing ships, the Sargasso Sea was frightening and mysterious. A berry-like weed grows in the Sargasso Sea, and because ships were often becalmed in these latitudes, the weed could be seen for days on end and led to the belief that it was a trap for flotsam and for old hulks that would become enmeshed in a solid mat of growth. These were the days when the ocean deeps were believed to house monsters which could demolish ships, or at the least rear their sea-serpent heads above deck level and pick off a few tasty sailors.

We know now that human beings do not form the staple diet of any mysterious denizens of the deep, but we have discovered that there is a regular cycle of life in the sea and with the rapidly increasing world population any improvements in this life cycle could be of inestimable benefit to mankind. Big fish eat little fish and the little fish live on even smaller ones, until at the 'beginning end' of the chain of life we find minute plant growth on which a drifting population of animal plankton lives. The one-celled drifting plants, known as phytoplankton, require similar sustenance to plants on land. They build up their carbohydrate cells by taking carbon dioxide from the sea-water, using at the same time minute quantities of dissolved salts containing such elements as potassium and phosphorous. Sunlight is essential for the breakdown of water to provide a supply of hydrogen and to make the carbon, hydrogen and oxygen combine with the metallic elements to form new living cell material.

The commonest of the small plants in the sea have a thin outer coating of silicon, a material which is of the same composition as sand and which is more resistant to dissolving than the shelly calcium carbonate used as shells by many marine animals. It is for this reason that the silica remains of these minute 'diatoms' are found in sediments in the deepest parts of the oceans. The diatoms form the 'grass' of the sea on which minute animals graze, and these grass-eating members of the ocean community are in turn eaten by small, shrimp-like carnivorous animals. There is a big loss at each stage of the food chain. It takes 100 pounds of plants to feed ten pounds of grass-eaters, and these only provide enough sustenance for one pound of shrimps. The process does not sound very efficient, but it works, and improvement may one day increase the total sea harvest.

The blue whales have short-circuited part of the normal cycle by making small shrimp-like animals called krill their staple diet. The whales manage to eat several tons of these a day (10,000 tons in a lifetime) by sucking in the krill-laden water and expelling the water through their teeth, which consist of a 'whalebone' filter on the roof of the mouth. The whalebone whales feed mainly in the southern summer on the large crops of krill that grow in the Antarctic waters where the upwelling currents occur. They grow up to 100 feet in length and weigh 150 tons, a size sufficient to nourish their calves, which may weigh two tons at birth and drink a ton of their mother's milk every day. During the last fifty years the whale population has decreased enormously due to over-fishing by the whaling fleets of the world. All nations except the U.S.S.R. and Japan have stopped their whaling operations, but the heavy catches of the last few decades, made in defiance of good marine biological advice, have reduced the whale population to such an extent that it will take a long time to recover.

There is presumably a surplus of krill in the Antarctic region and it is possible that it might prove a valuable source of food for man. Russian trawlers have demonstrated that catches of up to six tons of krill can be made in half an hour, and if human beings did not fancy the shrimp-like food direct, there are many ways of processing fish to produce an acceptable diet. Many oceanographers have tried eating even smaller marine life than krill by making soup of their plankton collections!

A second type of whale is more traditional in its feeding habits and lives on large sea animals. The sperm whale, of *Moby Dick* fame, is a great squid eater, and it is believed that big battles occur in the ocean depths between giant squid and large sperm whales. The whale has the advantage of powerful jaws and teeth but since he is a mammal and must come to the surface to breathe, he could conceivably be held under long enough to suffocate. Whales have been caught with tentacles of squid up to thirty feet long in their stomachs, while their bodies show scars of the large squid suckers.

The giant squids have two long, sinuous tentacles and eight shorter ones and are known to reach an overall length of fifty feet. Such an animal could pass for a sea-serpent and may well be one of the visual causes underlying the persistent reports of sea-serpents throughout nautical history. Although the squids, live hundreds of feet below the surface, it has been suggested that during a struggle with a sperm whale the surface threshings of the two fifty-foot animals could give the appearance of a giant serpent and put fear into anyone who did not realize what was happening.

There have been other 'candidates' in the search for the mythical serpent. For example, there *may* be eels 100 foot long; a young eel (or elver) five feet long was spotted in the sea. Since a young eel of three inches is reckoned to achieve a length of five feet when fully grown, it could be argued that the five-foot elver would eventually achieve a length of 100 feet. Perhaps this is the inhabitant of Loch Ness, although the latest observations of the murky Loch Ness water made by careful sonar surveys do suggest that there is nothing large beneath the surface apart from driftwood. However, Loch Ness, will still attract many observers and wild theories. One can always assume a monster with sufficient acumen to hide from the acoustic beam of the sonar; we are learning more every day about the audio capabilities of fish, especially those of the cetacean—or whale—family.

Another candidate for the sea monster is the oar-fish, which is long and flat and for this reason is sometimes known as the ribbon fish. This species is known to grow to a length of twenty-two feet and it is therefore possible that much longer examples may exist. To quote from Robert C. Miller, in *The Sea*:

> The giant oar fish has a bright red dorsal fin extending the whole length of the body from the head to the long, tapering tail. The forward end of this fin has long spines, which can be elevated into a rather tall red crest when the fish is disturbed. This would seem to take care at least of the tales of sea serpents with bright red hair!

On a more mundane note it has been said that a school of porpoises in a line would give a very good imitation of the traditional sea-serpent. However, the ocean is deep and large, and perhaps we can hope that in the warm central swirl of the Sargasso Sea there may be a colony of some giant sea animal which will substantiate all the sightings of the past centuries. Maybe some old plesiosaurs still exist. After all, it was a long time before the supposedly extinct coelocanth was discovered.

The fish harvest collected by man from the sea is about sixty million tons a year, which is equivalent to several thousand million pounds in monetary value. The world's largest fishery is off Peru, where the anchovy feeds in vast quantity on the large crop of phytoplankton resulting from the upwelling of the Humboldt Current. About ten million tons of anchovy are caught each year in this area; the catch is processed to make fish meal, which is a high-quality protein food for pigs and chickens, and after purifying can be made very acceptable to human beings in the form of fish flour.

There is some worry that the anchovy fisheries of the world will decline, as has the whale catch, due to over-fishing. However, the anchovy life-cycle is much shorter than that of the whale and so re-establishment of the population could be much more rapid. It is possible to determine the age of some fish by a microscopic examination of the fish's scales, for example, or by counting the annual rings in the calcareous concretions. Fish life-cycles can therefore be followed, and in particular the age of maturity may be found. In order to preserve a species of fish it is important that the young immature fish shall be allowed to grow, and the biological study of the sea provides the facts on which sensible restrictions to fishing practices can be based. As our knowledge increases, provided nations co-operate, it should be possible to gather increasing quantities of food from the sea without harming the standing population.

A recent substitute for fish meal which may help to bridge the protein gap in the coming years is protein manufactured from petroleum. Certain strains of yeast feed on some of the waxes contained in crude oil, and by adding oxygen from the air, and nitrogen in the form of ammonia, a continuous fermentation process can be established, with oil being fed in at one end and yeast coming out the other. The dried yeast has very similar nutritive properties to fish meal. As in the fish cycle of the oceans, some mineral salts such as potassium and phosphorus are needed as 'fertilizer' to the protein process.

It may appear odd at first sight to turn oil into something analogous to fish meal, but it is really only the completing of a biological cycle, petroleum being derived initially from decaying marine life. It takes millions of years for the oil to accumulate in quantities which can be produced commercially, but oil is being formed today, and traces of petroleum can be extracted from the shallow sea-bed muds in large river deltas. It has been suggested that the big oilfield areas of the world correspond to places where prolific animal life existed in past geological ages, such as the upwelling areas off Peru or Walvis Bay, but this is not certain, and it is possible that the large oil accumulations are the result of squeezing small concentrations of oil from large bodies of sediments.

Although this yeast from oil could provide millions of tons of good protein a year, the supply is limited, because oil, being a fossil fuel which was laid down millions of years ago, exists in finite quantity in the world. An estimated million million tons of recoverable oil exist on the earth (some oil is held in such fine porous rock that it will not flow to the surface). This will last some time, even at the present rapid rate of increased demand, but one day it must run out, because the rate of replenishment is in millions of years, while it is being used in tens of years. The protein from oil will provide an opportune stop-gap while other sources of food are being developed, or while population levels are reaching a more reasonable figure.

In this interim period the research into the supply of food from the sea may well increase the steady annual supply, although many marine biologists have warned that, even with the enormous volume of the sea, we may be near the maximum crop that can be supported. Already there are signs of falling off in output of salmon in both Atlantic and Pacific, of cod in the Barents Sea, and of Californian sardines.

If we are to progress from hunting what fish exist in the sea to the type of animal husbandry that we employ on land we must follow the examples provided by farmers over the past two hundred years. Immature fish must be preserved from the fishermen by restricting the mesh of the nets. Young fish must also be protected from animal predators. including their own kind; because, of the large number of fish larvae that are produced few grow to maturity—most are eaten by their parents and acquaintances at a very early and tender age. One way to increase the number of viable young fish is to hold them in tanks on land until they are large enough to fend for themselves. This type of experiment has been tried in Britain; it is difficult to assess the results of such operations, but it is obviously a step in the right direction. Another method of protecting the young is to provide shelter; in Japan old tramcars and other angular rubbish have been dumped on the sea-bed to provide cover for small fish. In north-east England amateur aqualung divers have been enlisted by Professor Bellamy to plant artificial seaweed, again to provide a forest of protection for young fish.

In addition to giving the young fish a protected start in life, some good results have been obtained by transplanting small fish from poor feeding grounds to such areas as the Dogger Bank, where ample food supplies allow growth at four times the normal rate. However, much of the available food in the sea and on the sea-bed is eaten by starfish, molluscs and crabs, which are not wanted by human beings. In many places the unwanted animals eat three-quarters of the available food, and therefore for successful fish farming it will be important to devise methods of removing these predators. This is especially true when fish cultivation takes place in enclosed areas, such as lakes, lochs or lagoons, which have to be supplied with adequate food for the fish. The farming of such fish as plaice is still in the experimental stage in Britain, but in Japan commercial prawn cultivation produces 100 tons a year for the Tokyo market. One of the factors permitting economic fish farming is the price of the product, which in Japan is high, whereas in Britain fish is still comparatively cheap.

Fishponds for carp and trout have been maintained in Europe since early monastic times, and in suitable places can still be economic. It is possible that sewage will be used increasingly as nutrient for this type of fish farming; in addition, an increased rate of growth will be made possible by using warm water from power stations. In Scotland sole and plaice grow to maturity in eight months instead of four years when brought up in the warm cooling-water effluent of an atomic power plant. In an age of ever-increasing population, calling in turn for greater power consumption, and producing ever growing quantities of waste, it is essential to integrate the living processes as much as possible.

Sheep and cattle have been improved by selective breeding to give better yields of wool and milk and to carve up into larger quantities of the kind of meat humans prefer. Selective breeding of fish is also feasible, and has been successfully employed in growing extra large trout in the U.S.A.

While the fish farmers are discovering ways of improving their yields in confined waters, others are finding out whether it is possible to duplicate Nature's methods in the open seas. The upwelling currents bringing nutrients to the shallow-water continental shelves provide the world with the few prolific areas of fish growth. It should be possible to make some artificial upwellings, especially in this present age of powerful civil engineering capabilities. Recently scientists from the Lamont Geological Observatory in the U.S.A. have made the preliminary tests of what may one day be a do-it-yourself upwelling. A 3½-inch pipe was used to siphon cold bottom water from a depth of 3,000 feet to feed rich nutrients from the sea-bed into small pools at the surface. This may prove to be a cheap way of feeding fish farms, but it may also be used on a large scale to flood the top layers of the sea with rich bottom water, and thus encourage valuable fisheries like the anchovy activity off Peru.

The plants of the sea do not consist entirely of the minute

phytoplankton which form the beginning of the marine animal food chain. Large plants, which although referred to as seaweeds, do exist along the shore and have been important to man for a long time. Seaweed was once the primary source of iodine, and has long been used as a fertilizer. It contains as much potassium as farmyard manure, although it only has one-third of the phosphorus content. In Wales, laver bread is made from seaweed or carageen moss, and extracts of seaweed form agar jelly which is used in the biological laboratory and also in a variety of foodstuffs. The ice-cream industry uses alginates from seaweed to give the smooth creamy taste to their product and alginates are used extensively in sauces and soups. It is an interesting thought that, if certain seaweeds can concentrate iodine from the extremely dilute condition in which it is found in sea-water, other seaweeds might be developed to gather further valuable elements from the sea.

The Gulf Stream is too warm to encourage the kind of fish which are the main catch of North Atlantic waters. It flows to the east of the great cod fisheries of the Newfoundland Bank, and it may perhaps help, in its cooler reaches, the fisheries off Iceland; but as Piccard noted during his submarine voyage in the Gulf Stream (see Chapter 6), there was an unexpected dearth of animal life, at least in that part of the Stream which runs from the Florida Straits to well east of Cape Hatteras. Although there are large meanders and eddies in a horizontal plane, there appear to be no local circulations which reach to the sea-bed, so that, with no mechanism to bring nutrients to the top few hundred feet of the sea, no great abundance of marine life is to be expected.

The Gulf Stream does, however, follow its normal crazy behaviour in changing the latitudinal *status quo* by bringing well developed specimens of tropical fish much farther to the north than they would normally venture. The Gulf Stream extends the range of many West Indian fish species far northward in summer months, so that there is a strong tropical flavour to the pelagic life of the mid-Atlantic. The Portuguese man-of-war, that curious jelly-fish with 'sails', is really a whole colony of specialized individuals drifting along like a floating beehive or ant-heap and with a nasty sting behind its elegant camouflage. It travels north on the warm current which streams out of the

Bimini plug-hole, sometimes to North European coasts. Other jelly-fish of a more familiar type are carried by the Gulf Stream to the shores of New England from Caribbean waters, much to the annoyance of the local swimming enthusiasts, but beautiful to those who appreciate the lovely colours and gentle pulsating movement of the many varieties of primitive sea-life.

The Grand Banks and the cold Labrador Current ensure that the warm sea-life comes to a halt at the boundaries of the Gulf Stream. Some of the cold-loving shellfish such as winkles and whelks have surmounted the barrier of cold and hot water to settle in New England, but the Gulf Stream south of Cape Hatteras is not well inhabited, probably because of the fast-moving current, which must make life difficult for bottom living animals. The southern part of the Gulf Stream can be a hunting ground for giant rays and other warm-water monsters of tropical waters. The tarpons, sailfish and bonefish, together with five-foot barracuda and big groupers, are all found in the Bimini area. A giant Mantua ray, twenty-two feet from wing to wing, and weighing many tons, was caught in the 1920s. This flat fish, with the shape of a skate, is quite a harmless creature, unless on the end of a hook and line, when he can tow a boat at a speed of five knots, and might well upset it in his flounderings. It is said that the giant ray will pull a ship out to sea if he gets entangled in its anchor chains.

Interesting observations have been made by Picket and Wilkerson of the U.S. Navy Oceanographic Office, when flying 500 feet above the Gulf Stream during routine patrols to measure the northern temperature boundary of the Stream using an infra-red radiometer (see Chapter 8).

> There is one thing we do know: the edge of the Gulf Stream, the interface between the flowing warm and the more stable cold Labrador water west and north of the Gulf Stream, is a very interesting place. The sea states in the two different bodies of water at any given time may be entirely different. On one occasion we noticed 8-to-10-foot waves in the Gulf Stream and relatively calm water beyond the interface. The colour difference, of course, is startling to behold. The Stream is a deep blue, whereas the ocean

> through which it flows is a murky grey-green. The interface is also a great collector. Repeatedly we have noticed large schools of fish swimming just inside the wall in the warm water, including tuna and blues. We have now taken to reporting these to Bureau of Commercial Fisheries. One time we spotted a herd of some 200 sea turtles, each as large as a wash tub, bucking the current swimming south—again just inside the warm water layer. Along this interface, one also finds thousands of birds. When we come upon them they are on the water, but the noise of the airplane sends them aloft. At one point, right on the interface we spotted a large baleen whale floating belly up, obviously dead. And lumber: on just one flight we must have spotted enough lumber to build a couple of skyscrapers (if skyscrapers were built of wood)—every stick of it caught just at the interface.

In the Atlantic, schools of medium-sized blue tuna may number a thousand, stacked in many layers. The catches of tuna in the Gulf Stream vary from one part of the year to another, and generally tend to move farther north in the summer. The yellow fish, which is the most sensitive to cold water, only goes north of the Gulf Stream in the warmest period. The big eye and the albacore are more widespread, while the bluefin is the most migratory of the whole species in this western Atlantic region. The bluefin spends the summer on the continental shelf between Newfoundland and Cape Hatteras and in winter departs towards the warmer water south of the Sargasso Sea. During the summer vacation the bluefin schools may be seen 'pushing' in a dense group, and making a wave like a motor boat. Sometimes the fish mill around in one place, and when feeding viciously they make large splashes at the surface and often leap clear of the water.

The biology of the warm deep central mass of the North Atlantic circulating water, which we know as the Sargasso Sea, is not very complicated. The seaweed itself is similar to the bladder seaweeds found on the shores of Florida and the West Indies, and it is possibly renewed from these sources by the Gulf Stream and the subsequent North Atlantic clockwise circulation. In 1492 Columbus drifted through the sargassum weed, and a crab living on a bunch of it was hauled on board his ship. Many centuries later the Danish expedition in the ship *Dana* spent a long time examining the Sargasso Sea and found many other creatures living in the area. These included a sea-horse, snails, shrimps, a flatworm and four species of fish as well as the crabs seen by Columbus. Some of the animals have developed camouflage colouring and shape to mimic the yellow sargassum weed.

If weed from Miami and the New England coast can reach the Sargasso Sea, it is probable that this central part of the North Atlantic circulation acts as a sink for some of the liquid waste that is poured out to sea by the present generation of sea polluters. In this respect it is interesting to note that the American nerve gas was dumped in the Florida Current. It is unlikely that any leak will occur, but if it does, any waste products will tend to go outwards towards the hub of the North Atlantic circulation rather than to the U.S. coast. It will be interesting to see whether there is any sign of oil spilt in the Atlantic accumulating in the Sargasso Sea.

The old bogy of an ocean desert carpeted by a thick layer of golden yellow weed, interspersed with the rotting hulks of old sailing ships, has now been laid, together with that other canard that kept going into the 1920s: namely that wrecked iron ships only sank to a limbo where the water was so dense, due to the pressure of the water above, that the iron would float. Even propeller-driven ships were supposed to end up in the Sargasso Sea, the weed being so tenacious that it would stop the screw rotating, leaving the steamer to drift into the island of lost ships with the others. The facts, as many oceanographers will bear witness, are that the Sargasso Sea is not really exciting at all. The patches of weed are often clumps of a few feet in size, with larger aggregations of tens or hundreds of feet, and the thickness is a few inches—certainly not safe to walk on. The water is smooth and a beautiful blue. Perhaps it really was a sink and the plug has been pulled to remove the debris of the last few centuries.

10

PASSENGERS IN THE GULF STREAM

There is another biological aspect to the Gulf Stream—namely that which is concerned with the effect of this ocean current on plants and animals found on land rather than in the sea itself. As we have seen, there are secondary effects of the Atlantic circulation which help to provide a warm climate for parts of north-west Europe; this unusual climate is demonstrated biologically by the existence of species of flora in unaccustomed latitudes and by birds which breed in places which at first sight appear to be well outside their normal habitat. There are, in addition, arrivals from the west which not only provide solid evidence of the transport capabilities of the Gulf Stream, but also link up with that formerly mysterious central limbo of the North Atlantic, the Sargasso Sea. There are two particular species, one animal, one vegetable, that are carried for thousands of miles by the Gulf Stream. Both of these passengers are, loosely speaking, seeds: the first to produce plants and trees, the second to provide one of the tastiest seafoods, the eel.

The sea beans which are washed up on the shores of south-west Ireland and north-west Scotland must have been almost as much a mystery to the local inhabitants as were the *cocos-de-mer* which floated on to the coast of India and the Maldives from the Seychelles Islands. The *coco-de-mer*, or double coconut, is a memorable object, both because of its large size and its sexual suggestions, and it is not surprising that it was regarded as the seed of a peculiar plant that grew underwater, hence the name. It was not a long step to credit the *coco-de-mer* with aphrodisiac properties and it was also used as a powerful antidote to poison, so that the beaches of western India and the Maldives were scoured to find these valuable gifts from the ocean. The Seychelles Islands were probably visited by Portuguese traders in the sixteenth century, but were not inhabited, except by visiting pirates, until the eighteenth century, when the secret of the double coconut was revealed. The giant 100-foot high primitive coconut palms grew originally only in the Val de

Mai on the island of Praslin, and can be seen today as male and female trees which fertilize and bear fruit, in a manner which might almost be a prototype experiment for human beings. General 'Chinese' Gordon of Khartoum fame was firmly of the opinion that the Seychelles were the original Garden of Eden.

It is not reported that the seeds washed up as tropical jetsam on the Irish coast were ever esteemed to have stimulating or medicinal properties, but they were supposed to originate in plants which grew underwater. The beans which arrive on the shores of Ireland and Scotland are conspicuous objects, which may exceed two inches and it was Sir Hans Sloane, whose bequest to the nation was partly responsible for the foundation of the British Museum, who identified some of the beans as being the seeds of plants indigenous to Jamaica. Once a tropical land source had been suggested, several other seeds were recognized, but there still remained the possibility that the objects had been carried by sailors as souvenirs of their visit to the other side of the Atlantic. The wide distribution of the seeds and their continuing appearance finally convinced the scientific world that there was a sea transport mechanism which could raft floating material from the Gulf of Mexico to north-west Europe. The mechanism is, of course, the Gulf Stream.

There are other tropical products carried by the Gulf Stream to the shores of Ireland, such as 'blue snails' and Portuguese man-of-war. These passengers from the New World have been arriving for some time, as evidenced by the fossilized sea beans found in a low-lying bog in south Sweden. It is interesting to note that the sea beans have a hard coat which preserves them from the sea-water, so that they are capable of germination on reaching the shores of Europe. One seed was identified by the naturalist Sir Joseph Banks from a plant which was successfully grown from a sea bean washed up on the west coast of Ireland. Some of the indigenous wild flowers of the west Ireland coastal fringe have origins in America, and almost certainly have been transplanted with the favour of the Gulf Stream. For example, the American water plant, the 'slender naiad', is found near Roundstone and Slyne Head, and as R. L. Praeger says in *The Way that I Went*:

> Their special interest, apart from the great beauty of

some of them lies in this—that they are rare in or more often absent from Great Britain and the central parts of the Continent, and that they belong, not to the multitude of plants which migrated from middle Europe across England to Ireland in past times and have settled down over both islands, but either to countries far to the south of Ireland—Spain and the Mediterranean region—or to distant North America. Research has proved abundantly that they are not human introductions, but have migrated by natural means. These groups have their Irish headquarters in Kerry, or in Connemara, whence some species continue northward especially to Donegal. A few of the southern plants have colonized south-western England on their way from the Spanish area to Ireland; while a few of the Americans are found sparingly in Scotland, or north-western Europe. What are these plants? One of the finest of them is the Strawberry-tree, Arbutus, which gives character to the Killarney woods by its glossy leaves, Lily of the Valley blossoms, and prickly scarlet fruit. Then there is the Saxifrage which till lately was considered to be the 'London Pride' of English gardens, but is now known to be an allied plant found in the Spanish peninsula, spread along the Irish western and southern coasts and also in Wicklow; and the Kidney-leaved Saxifrage, which attains its main Irish development in the south-west. There is the Large-flowered Butterwort, which covers the hills and bogs of Kerry and West Cork, bearing a rosette of shining yellow leaves and in May a cluster of great deep purple violet-like flowers. The Irish Spurge, too, so common in the south-west and spread sparingly as far as Donegal, is a southerner. And there are other southern plants in western Ireland which do not grow in Kerry, but farther north, the most striking being three species of Heath which are referred to in the pages dealing with Connemara. The American ingredients in the Irish flora are fewer and not so conspicuous, but they include a highly fragrant Orchid—a species of Lady's Tresses, found in Europe only in south west Ireland—and another closely allied to it, confined to the Lough Neagh basin and a single place in west Scotland. Another remarkable plant is the Pipewort, spread from Cork to Donegal, occurring also in west Scotland, and next in North America; and there is the Canadian Blue-eyed Grass—an iris rather than a grass—widespread in the west of Ireland. It is specially significant that the members of these Hiberno-Iberian and Hiberno-American groups are not altogether limited to the vegetable kingdom; some animals show just the same peculiar distribution. Perhaps the most famous is the Spotted Slug of Kerry, elsewhere confined to Spain and Portugal: and there are a number of others, including molluscs, beetles, false-scorpions, wood-lice and earthworms, found in Ireland, and elsewhere only in the Pyrennean or Mediterranean regions. One animal form, a freshwater sponge, widespread in Ireland and known also from western Scotland, belongs elsewhere exclusively to North America, and reinforces the American group. One could fill a book with discussion of the many problems raised by the presence in Ireland of these animals and plants—the date of their arrival, the manner and the route by which they travelled, and the meaning of their curious distribution: but here we may be content with a recognition of their presence and of their peculiarities.

One must remember that, over a long period, the number of rare occurrences becomes significantly high, so that quite unusual theories may be postulated to account for some of the odd things that we discover. The Gulf Stream surface transport system is certainly an important factor in the migration of plants from America to Europe.

The Gulf Stream is also the supplier of the eels of Europe. Although eels look like snakes they bear no relationship to reptiles and are fish specially adapted to their chosen manner of life and their means of locomotion. There are eels everywhere, from inland freshwaters to the depths of the oceans, but it is the so-called freshwater eel that is of interest in connection with the Gulf Stream. This 'common' eel differs from the rest of the eel family in having small scales, which are placed in groups, those in one group running at right angles to those adjacent. The eel has a large mouth with strong teeth as befits a rough, tough predator, who devours any meat, dead or alive. The well-known conger eel does not have scales, and its dorsal fin starts nearer the head than in the common eel; it spends its whole life in the sea, but is also edible. Congers are generally longer than freshwater eels, and may grow to more than eight feet in length with a weight of 120 pounds. The eel has a most curious life-history, unknown to many of those who enjoy jellied or smoked eel, and those eel connoisseurs should occasionally give thanks to the Gulf Stream when indulging in their favourite seafood. Although eels live in freshwater ponds and rivers on land, they breed in the sea. This is the opposite way round to salmon, who lay their eggs and grow up in fresh water and travel out to sea to live their adult life. The eel starts its life in deep water in the Sargasso Sea to the east of Bermuda. The young do not look much like eels, being flat, translucent leaf-like creatures, with pinheads and large bodies. Formerly these young eels were classed as a fish and were called leptocephali. According to Dr. D. W. Tucker, after a year or more the leptocephali thin out and become worm-like, and at this stage they make for land and fresh water. One-year-old leptocephali reach the North American coast and settle in the rivers and lakes to grow into three-foot-long eels. Three-year-olds take up their abode in Ireland, while the eels that go to Scandinavia and the Mediterranean may be as old as five years when they arrive. It is almost certain that the migration from the Sargasso Sea takes place by a process of drifting in the Gulf Stream, with a slow progress determined by the meandering flow of this ocean current, rather than by a mainstream circulation, which should only take about one year for a complete circuit. The young eels, or elvers, soon gain dark pigments on arriving in muddy offshore waters, and their transparent appearance, which is a useful camouflage in the open sea, finally disappears. The eels grow rapidly, in fresh water, and when they feel the urge they migrate back to sea to breed again in the warm quiet waters of the sub-tropical Atlantic.

A saltwater fish which spawns in fresh water like the salmon is described as anadromous; the eel, being the opposite in both senses to the salmon is described as catadromous. The European and American eels ('Anguilla' is the proper name) are very alike and spawn in adjacent areas of the open Atlantic, to the south-east of Bermuda. There is a certain amount of argument about the difference in location of the U.S. and European breeding grounds, but there is no doubt that both groups of eels start their lives as eggs a thousand feet or more below the sea surface in the Sargasso Sea area. The young eels are well adapted both to drifting and to survival, with their leaf-like flattened form, and their transparent appearance. Presumably the density is adjusted with growth, so that the young larvae float to the surface layers when ready for their ride on the Gulf Stream. The observational evidence is that the leptocephali travel near the surface of the Gulf Stream drift; perhaps the larger proportion of those destined for Europe take the slower route provided by the clockwise-rotating North Atlantic circulation, rather than by the faster moving periphery which we call the Gulf Stream.

Eels are secretive animals who like darkness. Their sex life is shrouded in the ocean deeps where there is very little light, and even in their migration from their freshwater lifetime habitat they prefer the dark nights to travel. The young are hatched out in May and June and drift upwards towards the surface light to latch on to the Gulf Stream 'transport'. When they near shore they gather in large numbers to await a spring tide which will help them to move upriver into lakes and land-locked ponds. The young elvers are agile and can travel quite large distances overland. There is a story of the Irishman who took a short cut across a field after leaving the local pub, and was aghast to see hordes of snakes wriggling through the long grass. This was against all the canons of St. Patrick, who banished snakes from Ireland—but of course these snakes were eels.

The small eels carried by the Gulf Stream reach most of the rivers and ponds of north-west Europe. Some of the largest eel populations are found in Ireland. For example, twenty to thirty million elvers travel each year up the River Bann alone. These elvers live from seven to fourteen years in Lough Neagh and other inland waters and grow into sleek fat eels, which may be over three feet long if female. The males grow to only half this length. The eels like to burrow in the soft mud at the bed of the pond during the day, making stealthy excursions at night to feed. After their early life in fresh water the eels feel the irresistible urge to breed, and move in droves down to the sea. This is their last journey, and in preparing for it the eels put on a thick layer of fat. This is their food reserve for the long journey to the Sargasso Sea, since they do not eat again after leaving the fresh water; they can no longer eat because their digestive tract has shrivelled up. The body changes from a yellowish colour to a silvery black, and for a long time yellow and silver eels were thought to belong to different species.

The life story of the eel was elucidated before the First World War by Dr. Johannes Schmidt, a Danish biologist. Dr. Schmidt tracked the eels back from deep water in the region of the Faroes Islands to the Atlantic, and found that the young eels became smaller as he went farther west, and from being worm-like when near the European shores, changed to the transparent leaf-form of leptocephali.

The silver eels make their way out to the sea, and probably breast the current of the Gulf Stream in order to reach their breeding place to the south of the Sargasso Sea. If they were wise they would probably take advantage of the clockwise circulation of the North Atlantic waters and arrive at their destination by going first south then west along the Equatorial current. Perhaps the urge to get to their birthplace is so great that they go by the direct route in spite of adverse currents; perhaps their store of fat is not large enough and the eel contingent from northern Europe never does arrive. There is still a certain amount of argument on these points.

The examination of eels in the aquarium suggests that eels breed only once in a lifetime. Both male and female prepare themselves by accumulating a reserve of fat, and then they stop feeding for several months while the male makes his milt or sperm, and the female develops large ovaries. The lack of food causes loss of sight, and the skin becomes ulcerated. The bones become soft due to lack of calcium, and the teeth disappear. Presumably a similar state of affairs occurs on the journey across the Atlantic. In the aquarium the female dies before the eggs are properly ripe for release, and it has been suggested that the pressure of deep water is needed to force the eggs out. This fits in with the observation that the young larvae first appear at depths around 1,000 feet in the seas to the south-east of Bermuda.

Johan Schmidt followed the small eel larvae back to their breeding ground, and in the course of his studies discovered that the American eel and the European eel breed in overlapping areas of the Sargasso Sea, which formed the basis of Dr.

Tucker's theories (see p. 123). The common source, on this theory, is America, the long journey from Europe being so arduous as to defeat any successful migration from that area. It is a sad thought that all these eels, having committed themselves to a self-imposed six-month fast, never reach their breeding grounds. On the other hand, it does make for a clearer conscience for those who catch the migrating silver eels in millions, and also for those who delight in eating them.

Eels do not swim very fast, a top speed of about 2½ miles per hour being reported, with a cruising rate of between five and ten miles a day. The record is 32½ miles a day, but was only kept up for two days. The European eel has to go 3,500 miles, while his cousins from the Black Sea must travel over 5,000 miles to reach the Sargasso Sea breeding grounds. At five to ten miles a day this would entail a journey of over a year; but the eel has a store of fat good only for about six months, and so, although he sets out from fresh water with the best of intentions, he probably finishes up not very far from his starting-point. The rate of progress will slow up as strength ebbs, and the burning up of the energy stored as fat will be more rapid at sea than in the comparative calm of an aquarium.

Those who oppose Dr. Tucker's belief, and assume separate American, European and Black Sea varieties of eel, together with separate breeding grounds, use the example of the salmon to demonstrate the tenacity of fish who are determined to reach their traditional breeding ground. Certainly salmon travel great distances; they move upstream to their shallow pools without taking food, and lay their eggs when in a sorry, emaciated state. The longest recorded journey for an Atlantic salmon is 1,730 miles in 328 days, which is much less than the eel would need to do, yet the salmon travels more quickly than the eel, judging from checks on salmon that have been marked and subsequently caught. It is odd that eels are not caught by fishermen in the Atlantic; it is possible that they swim too deep, but it could be that they have tired in their fruitless efforts and have been eaten by another predator in the eternal fish cycle. An argument against Dr. Tucker's hypothesis is that natural selection would, in the course of many generations, discriminate against those American eels which bred in the area which led to Gulf Stream circulation to

Europe, because these eels never had any progeny. On the other hand, it is possible that the vagaries of the North Atlantic circulation allow some leptocephali to be swept towards the American shore, while others drift on for several years more before reaching Europe and the Mediterranean.

The conger eel uses breeding grounds between Gibraltar and the Azores, and also in the Mediterranean as well as sharing the Sargasso Sea area with the freshwater eel. It is surprising that the European eel has not found some breeding ground nearer its freshwater habitat; perhaps it has, and we have not yet discovered it. It may be, however, that the story of the fateful migration is the true one and the annual European crop of eels is of American origin.

The eel has been a favourite food throughout historical times. Eels were popularly believed to be spontaneously generated in mud, a natural supposition which explained their existence in landlocked lakes together with no visible spawning or producing of young. In medieval times eels were sold both fresh and salted in the Bann valley of northern Ireland. Today most of the enormous catch from Ireland is despatched to England and north-west Europe where the eel is still considered a delicacy. The Bann valley leads to Lough Neagh, which is said to be rivalled only by the Dutch Ijssel Meer as a feeding ground. The value of the eel harvest in Ireland is about a quarter of a million pounds sterling a year, and comprises a weight of nearly a million pounds of eel. About two-thirds of this is caught by fishermen in Lough Neagh, the rest in traps in the rivers and streams. The eels caught in traps are all silver eels, since they are migrating towards their breeding grounds. About half the total catch are brown (or yellow) eels, since in fishing Lough Neagh it is not possible to separate those who are living their freshwater life and those who are preparing to migrate.

The season lasts from June to January, and it is the migrating eels which are caught in the traps and weirs. The streams are diverted into the eel traps, which may consist of a hut with a slatted floor, or nets used to collect the eels at suitably sited weirs. In the ancient days spears and 'eel rakes' were used but these are banned today and are used only by poachers. The eels do not like moonlight, and prefer to make their run when it is dark and stormy, and if possible with an onshore wind raising the tide. The nets leave a gap of one-tenth the width of the river so that a proportion of the eels can pass to the sea. This may be unnecessary if the young eels in the Sargasso Sea come from American parents, but it is a wise precaution where we are not absolutely sure of the eel's life-history. On good nights ten tons of eels may be caught, and the record in the River Bann is more than thirty tons in one night.

Even if we are not yet sure what happens to the migrating 'silver' eels when they leave the shores of Ireland, we do know that the continuing annual source of young elvers is provided with the help of the Gulf Stream, just as the sea beans drift across from the West Indies to western Ireland. There are other biological observations which reflect upon the presence of the Gulf Stream, albeit in an indirect way. During the present century the behaviour of the bird population points to a general amelioration of the climate of the North Atlantic. There has been a move northwards of several species of birds. For example, in Ireland the red-necked phalarope and the red-throated diver, which were reported to breed in Ireland in the nineteenth century, have withdrawn, and are assumed to be nesting further north, presumably because the fish they prefer have also migrated to water of a suitable cold temperature. The little auk has almost deserted Iceland as a breeding ground, because it gets its food from the edge of the Polar pack-ice, which has moved northwards. The commonest wild swan in Ireland was formerly the Bewick swan, which wintered in Ireland and bred in the Arctic lands of Russia and Siberia. Its place has been taken by the whooper swan, which breeds in the northern waters of Europe, Asia and Iceland. On the other hand the tufted duck, the turtle dove and the carrion crow, who like the milder climate, have moved in to Ireland. The changing bird distribution and the withdrawal to the north of the Arctic species can be seen in Iceland as well as in Ireland.

No one has produced a definite explanation for climatic fluctuations, just as no certain reason for the onset of Ice Ages and their disappearance has been forthcoming. There is clearly a fairly delicate balance of heat from the sun, affected by the composition of the atmosphere, and the heat absorbed by the oceans, and the heat exchange between the sea and the atmosphere. Such currents as the Gulf Stream must play their part in the heat transfer, and could possibly affect quite large areas of the earth if for any reason they changed course, or in a large measure altered their rate of flow. Just as we still do not know where the eels go, so we must make more measurements to find out how the world's climate operates.

11

EXPLORATION AND CONQUEST

The Gulf Stream has played an important rôle in the history of the New World. Before modern times—when engineering capability has become so great that inventors can talk of altering the Gulf Stream flow—the current pattern was one of the controlling natural features of voyaging across the Atlantic. In the days of sailing ships currents of a few knots could make a very real difference to the duration of a voyage. The clockwise circulation of the North Atlantic surface waters enables the voyage from northern Europe to North America to be made by a direct northerly route, breasting the adverse flow of the North Atlantic Drift, or the ship can travel first south, then west in the North Equatorial region, finally following the clockwise currents northwards past Florida to the American mainland. Both these routes played their part in the development of North America, and other aspects of the current pattern have affected the history of the Central American lands and the islands of the Gulf of Mexico and the Caribbean.

It is possible that the North Atlantic current pattern first made its impact on mankind in the earliest times. Plato describes the beliefs of Egyptian priests in the existence of a large island, greater in size than Asia Minor and Libya put together, and situated outside the Straits of Gibraltar. This is the earliest report of what has become the legendary island of Atlantis. Plato's informants further described the destruction of this large and prosperous island when it was overwhelmed by the sea. In medieval times many small islands both in the Mediterranean and along the western coast of Europe were claimed as the remnants of the lost empire of Atlantis; but the myth may have originated in a successful early voyage across the Atlantic to the Americas or to some of the fringe islands. Thor Heyerdahl has demonstrated that a boat or sailing raft built of papyrus and other materials available to the ancient Egyptians could have followed the currents and made a successful crossing. The return journey could also be made along the western and northern part of the Gulf Stream, with a final return to the Straits of Gibraltar assisted by the southward drift across the Bay of Biscay. The possibility of this return journey does not need the demonstration by papyrus raft, since a continuous supply of flotsam, in the form of beans and other seeds, is washed up on the coasts of Europe and is identified as originating in the West Indies (see Chapter 10).

The ancient voyages of discovery by the Egyptians may have been muddled with other historical events in the days when reports were passed down by word of mouth rather than in a written record. Modern archaeological research indicates that some Greek islands have suffered inundation by the sea, and sunken cities are now being investigated with the important diving aid of the aqualung. The Mediterranean area is one of great volcanic activity, and explosive eruptions can bury cities such as Pompeii overnight. Earthquakes and associated tidal waves can cause large slices of land to submerge and destroy cities in low-lying positions. The legend of Atlantis is, then, probably a garbled description of a large earthquake and volcanic catastrophe, which destroyed a well developed civilization. The account of the catastrophe became embellished in the telling of the story by placing the island outside the Pillars of Hercules, as the Straits of Gibraltar were called. This would make the story more plausible to audiences who had explored the Mediterranean quite thoroughly and had found no trace of Atlantis, and it would have a faint justification if there had been old reports of successful Egyptian voyages into the Atlantic.

Those who doubt that Columbus discovered America are flying in the face of well-entrenched traditional opinion, but there is a strong modern body of opinion which supports a 'diffusionist' theory of population of the Americas. The diffusionists believe that the Americas were populated well before Columbus' celebrated voyage, by sporadic arrivals, possibly by accident, of groups of people from Europe, Asia and Africa. Some of the evidence for this belief is based on a collection of old terracotta heads found in Central America and examined by Alexander von Wuthenau (Professor of the History of Art at the University of the Americas in Mexico). These old heads portray many different types—Negroes, whites reminiscent of ancient Greece, and Asiatic types, and could be accounted for by the arrival of visitors from other parts of the globe centuries before the New World was discovered by Columbus in 1492. The route which was probably followed by humans in migrating from Asia to the American continent is the shortest oversea path across the Bering Straits. This was undoubtedly used and accounts for the Eskimo and Red Indian Asiatic types. It is, however, a long way round for Negro and for Mediterranean migrants, and it is easier to believe, with Professor Wuthenau, that the longer routes across the central Atlantic from Africa and Europe, and across the Pacific from South-east Asia,were also occasionally used.

We know that the current pattern of the oceans has been the same as it is today throughout historical time, and therefore chance drifting across the oceans would have been possible during the past few thousand years. One old story, which has recently been substantiated by new evidence, is that of a Phoenician voyage in the seventh century B.C., which is recorded on a stone tablet discovered in Brazil in 1872. The Phoenician inscription describes an epic journey from the Gulf of Aquaba out of the Red Sea and around the Cape of Good Hope into the Atlantic. The ten trading vessels were intending to follow the coast round to the Straits of Gibraltar and thence back home to Sidon, but they were blown away from land in a storm, and presumably carried by the South Equatorial Current to Brazil. It is conceivable that the return journey could have been made by picking up the northern area of the South Equatorial Current, and following this into the Caribbean and then joining up with the Gulf Stream. However, the evidence for this voyage begins and ends with the carved stone record from Brazil and even this was believed to be an elaborate hoax when its discovery was announced a hundred years ago. The original stone was never located by the museum authorities who translated the record, but recently the old scrapbook of an American bibliographer, Wilberforce Eames, was found, containing a letter from the museum director in 1874, together with the tracing of the original copy of the Phoenician carved text. Experts believe that the style of the old Brazil inscription is genuine, and could not be the work of a nineteenth-century forger. Since the current pattern of the Atlantic Ocean makes it possible for the voyage to have been made, it is probable that the Phoenicians were early visitors to South America. However, the more cautious may be reminded of that notorious hoax of the early years of this century—the Piltdown Skull.

Some of those who wish to discredit Columbus do not go back to ancient times, but base their beliefs in early discovery on the Vikings. It is supposed, once again on oral evidence, this time in the form of traditional house sagas passed down from one generation to the next, that around 1000 A.D. Leif Ericson was blown off course on the way from Scandinavia to Greenland and discovered new lands to the south-west. Leif's father, Eric the Red, had settled in Greenland, the climate being milder than it is today, and Iceland was also inhabited by the Viking people from Denmark. The land to the south-west of Greenland was named Vinland, on account of the wild vines which were growing in quantity. The old sagas also report the existence of self-sown wheatfields, and an encounter with some swarthy, ill-looking local inhabitants who were paddling skin canoes.

A few years later, an Icelander, Thorfinn Karlsefni, after visiting the Ericson colony at Greenland, set out for the new Vinland with 160 followers. The journey was successful and the explorers settled for three years, living off the natural food and grazing that Nature provided. The story relates that there was no snow, so either the winters were warmer than now, or the settlement was a fair way down the North American coast. The new colonists wound up their operation in 1006, because the natives were becoming hostile, but the discovery of North America was blessed by the birth of a son to Thorfinn and his wife Gudrid.

As so often happens when old reports are handed down over the years, and not recorded on stone or paper at an early stage, the details of the early Viking landings are questionable. Some insist that the Vinland was Nova Scotia or Newfoundland, while others prefer Rhode Island as the first landfall. There is also argument as to whether the natives were Eskimos or Red Indians, and it is possible that the 'wine-berries' that gave the new territory its name were wild cranberries rather than grapes. There is little doubt, however, that several visits were made from Greenland and Iceland to what is now Canada and the U.S.A. in the eleventh and twelfth centuries. A period of cold which caused the Greenland settlements to be abandoned probably stopped subsequent colonization from the north. The Icelandic annals quote Bishop Eric of Greenland as going in search of Vinland in 1121, but since there is no further report of his voyage, and a new bishop was appointed in 1123, it must be assumed that the expedition was unsuccessful, or if one prefers to be optimistic, that the bishop liked the warmer living conditions of America and stayed there.

The next recorded voyage to the Americas is the well documented and historically famous landfall of Christopher Columbus, who reached an island of the West Indies on 12 October 1492. It is interesting that the southern route across the North Atlantic had not been attempted before but the reason could be that the Mediterranean was the thriving centre of the universe in medieval times, and that voyages to Asia took place overland or across the Indian Ocean. It is probable that Columbus discovered America under the misapprehension that he was going to India in a westerly direction from Spain instead of the previously used route to the east, but he must be given full credit for thinking up the idea of looking for land to the west, and for being so persistent with royalty, government officials and merchants that he raised the ships and finance needed for a sizeable undertaking.

Columbus studied old charts, and listened to reports of seafaring men, who must have travelled half-way across the Atlantic when visiting the Azores, and catching whales in the vicinity of these islands. It is probable that Columbus knew of the old reports of the Norsemen, and he would certainly be aware of the sea beans and other material transported by the ocean currents from some hypothetical western land. He subscribed to the belief in a spherical earth, and therefore deduced correctly that India could be reached by going west about the globe as well as eastwards. He thought that Asia was much larger than it is, and therefore did not place any land between Europe and Asia on the Atlantic Ocean side, so it is not surprising that when he finally made his American landfall he called the islands the West Indies.

Columbus began his voyage to the west on 3 August, so he made the crossing in about ten weeks. In addition to the main ship, the *Santa Maria* of 100 tons, two smaller craft, the *Pinta* and the *Nina* made up the small fleet, which carried eighty-eight men. The clockwise swirl of the North Atlantic current system obviously helped progress, and the ships passed the Sargasso Sea, where Columbus made some of the earliest recorded observations on the natural life. Columbus' report suggests that the southern route, as opposed to the cold Iceland-Greenland Atlantic crossing, was warm and like spring in Andalusia. However, as the weeks went by, the crew became restive, but their resolute admiral kept them going, and finally on 11 October the *Pinta* recovered some pieces of wood which suggested human work. Columbus pointed out a light that evening and land was sighted early the following morning.

Columbus made three subsequent voyages to the West Indies, and he and his lieutenants did a great deal to map and colonize the islands. On the fourth voyage he was forced to run his ships aground in Jamaica, because of their rotting timbers, in the small inlet now called Don Christopher's Cove. The Spanish Government encouraged the gathering of wealth from the islands, and began the slave labour tradition of the area by using the local natives for work in Europe and in the island plantations. British, Dutch and French seamen soon followed the Spanish and Portuguese into the New World, and for some centuries this warm and pleasant Gulf area was the scene of piracy, naval battles, native uprisings and traitorous behaviour, which does not seem to have ended.

It was during the early days of discovery of the West Indies, including the large islands of Cuba, Haiti and Jamaica (as well as the many smaller units that make up the Bahamas and the rest of the Antilles), that the Gulf Stream was first reported. Juan Ponce de León was one of Columbus' men on the second voyage, and had been appointed governor of Puerto Rico in 1509. During his stay in the West Indies, Ponce de León heard of a famous spring, known as the 'fountain of youth', which was situated on a fabulous island called Bimini. He was determined to locate this source of health-giving water, and in the course of so doing discovered Florida. On the way from Cape Canaveral (now Cape Kennedy) to the Tortugas Islands the Gulf Stream was encountered where it is at its most obvious —in the Florida Straits. Whether the fast-flowing Gulf Stream current had led to the myth of the fountain of youth, or whether some hot springs associated with volcanic activity were the source of the reports is not known. Ponce de León obtained permission from the Spanish Government to colonize some of the islands, and returned in 1521; but he clearly missed the elixir of youth, since he died in Cuba after being wounded in an attack by Indians.

With the advent of frequent voyaging between the Old World and the New, the Gulf Stream came into its own as the most comfortable homeward route from the West Indies to Europe. The weather can be bad on the home run across the northern Atlantic, but it is the way to make a quick voyage, taking advantage of the steady bonus of twenty to forty miles a day provided by the current. In a similar way the southern part of the Atlantic circulation using the Canaries Current and the Equatorial Drift made a pleasant routine voyage out to the new lands.

It did not take long before the new explorers moved to the mainland and discovered the gold and silver which had been amassed by the old civilizations of Central America. One of the most famous of the Spanish pioneers of these times, and one who matched his government's double-dealing with similar low cunning and duplicity, but withal, great courage and endurance, was Hernando Cortes, the conqueror of Mexico. Cortes was only nineteen when he went out to the West Indies in 1504, and he impressed the local rulers of San Domingo by his amiable manners and his military skill. Their judgment was to be borne out by the actions of the next few years, which began when Cortes joined Diego Velásquez on an expedition to Cuba. In November 1518, Cortes was sent from Cuba with ten vessels, six hundred men, sixteen horsemen and a few cannon to subdue the newly discovered Mexico. Velásquez changed his mind about the appointment of Cortes soon after the force had set sail and ordered his arrest, but was foiled in this particular piece of meanness by the troops, who admired their commander.

The combination of the Spanish ships, with their many-storied, fort-like structures fore and aft, together with the noise of the cannon, and the sight of men in armour on horseback, led to advance reports amongst the native Mexicans that the Spaniards were gods. Presents and ambassadors arrived to pay respect to Cortes, who soon founded the settlement which became the city of Vera Cruz, at the same time burning the ships as a message to the troops that there was no return unless they conquered the new country. It was soon discovered that the emperor and leader of the various rival factions in Mexico was Montezuma, and by careful playing off of one group against the other, Cortes raised a local army of several hundred Indians and with some three hundred of his own men visited Montezuma at Mexico with a view to taking over some of the riches of his empire. In the meantime, a little underhand work by Montezuma himself resulted in the killing of seven men of the Vera Cruz garrison and the myth of the invincible Spanish gods was destroyed. Cortes responded by capturing Montezuma, and burning alive those who had attacked the Spaniards. Montezuma was given a semblance of freedom in exchange for a large ransom in gold and precious stones and the acknowledgement of the sovereignty of the king of Spain.

In keeping with the general behaviour of the period, Velásquez at this juncture sent a force to make Cortes renounce his command, but the great soldier Cortes not only defeated the expeditionary force sent for him, but enlisted many of its members to his colours. Meanwhile the Mexicans had revolted against Montezuma and forced the Spaniards to retire, losing all their rearguard in a bravely fought retreat. But the military genius of Cortes reconquered Mexico and finally, in 1522, he was officially appointed governor by Charles V and the Pope. Cortes continued with his armed suppression of any opposition. Eventually, however, opposition to him mounted, both in Spain and in Mexico. His retainers were imprisoned and his policies thwarted. In 1534 Antonio de Mendoza was appointed viceroy and Cortes' power dwindled. He sailed to Spain for a brief visit, but was never to return, dying in Seville in 1547 while awaiting a ship to take him back to his beloved Mexico. The Church moved into Mexico bringing with it the Inquisition while the British Navy, together with a fringe of pirate fraternity, took over many of the West Indian islands.

The foundation of British naval supremacy, which existed from the sixteenth to the nineteenth century—with a few temporary setbacks—was laid in silence, in that there is no written report of the great seafaring voyages of the traders of the fifteenth century who exhibited a spirit of enterprise and who rapidly followed up discoveries of new lands. The English merchants sent out ships from Bristol and only hints of their achievements can be gleaned from the writings of Robert Fabyan, John Stow and William Wordsworth, as well as the sixteenth-century writings of Richard Hakluyt. Thylde deserves mention since he attempted the Atlantic crossing in the face of the gales of the 40°N region. Thylde was twelve years ahead of Columbus and was reported to be the most scientific seaman of his time. It was probably the Gulf Stream rather than navigational ability or seamanship which decided that the credit of discovering the Americas should go to the Mediterranean countries. If there had been a current in reach of the English Channel which carried ships to the west there would have been a different pattern of exploration of the Americas. The English explorers of the late fifteenth century were looking for the 'island of Brazil' twenty years before Columbus made his epic voyage. If only the Bristol captains had attempted to run south before crossing the Atlantic they would have enjoyed the same good weather that helped Columbus on his way. Not only can one drift across the ocean on the North Equatorial Current, but the trade winds are also favourable, while in the northern latitudes which are the sailing grounds of Scandinavan and British seamen, both the winds and the currents are adverse.

King Henry VII and the Bristol merchants were rewarded in 1497 when the *Matthew*, with its captain, John Cabot, a Genoese financed by Britain, made the Newfoundland coast via the northern route. The importance of Cabot's voyage, however, lies in its successful landfall rather than epitomizing the individual explorer. During the next few years the northern portion of the New England coast was visited by many ships. The chronicler Richard Hakluyt tells of a Mr. William Haskins of Plymouth who made three voyages to Brazil from 1530 onwards.

Other ships' captains chose the southern route across the Atlantic; typical of these was John Hawkins, who went first to the African coast and thence along the North Equatorial Current to the West Indies with a cargo of slaves. The savage Spanish exploitation of the local population had inevitably called for reinforcements of labour, and the black man from Western Africa was available and was traded for European goods both by the Arab infiltrations from the North and by the tribal chiefs themselves. Hawkins had a rough time at Vera Cruz when the Spanish treacherously attacked his ships after a friendly visit to trade. Some of Hawkins' men, survivors from one of his three ships, were landed in Mexico, but suffered torture and death at the hands of the Spanish Inquisition. This did not endear the Spaniards to Hawkins or to his friend Francis Drake, and the Spaniards later got what they deserved in the epic of the Spanish Armada.

Drake played his part in capturing from the Spaniards mule-trains laden with gold and silver while crossing the narrow Isthmus of Darien in the Gulf of Mexico. John Oxenham, another of the same English West Country 'gang', decided that the interception of the Spanish riches could best be accomplished by taking the Spanish supply ships in the Pacific Ocean, before they transhipped via the mule-trains to the Atlantic fleet. So he built a 45-foot pinnace, launched it in the Pacific Ocean, and captured two Spanish treasure ships. He made the unusual but pleasant gesture of allowing the enemy crews to depart so that the alarm was given and the gallant and generous Oxenham was captured when crossing the isthmus back to the Atlantic. The same mercy was not of course shown by the Spaniards.

Antonio de Alaminos, who had been with Columbus on the fourth voyage and who had sailed with Ponce de León across the Gulf Stream in the Florida Straits, was Cortes' commander of the fleet in Mexico. When sent on despatch duty to Spain, he quietly slipped out, to avoid the pirates and his own envious rivals in the vicinity, through the route north of Cuba and past Bimini into the fast Atlantic Gulf Stream flow.

It was at this time, in the mid-sixteenth century, when the New England territories were being developed, that the Gulf Stream probably played its most important naval and military rôle. It may even be imagined that the Gulf Stream determined the character of the two parts of the American nation. We have seen how discovery has alternated in using these two main routes across the North Atlantic to America over historical times. The explorer could either travel north of the North Atlantic Drift and have the cold winds and rough seas to make a 2,500 mile journey, or he could idle round the Canaries and follow the Equatorial Drift to the Caribbean, and then ride the Gulf Stream to the southern colonies of North America. The southern route was warm and comfortable, although possibly liable to becalming in the Doldrums if navigation was careless; the northern straight approach, undoubtedly favoured by the practical, hard-hitting inhabitants of northern Britain and Scandinavia, was tough and cold but would have the reward of plentiful fishing on the Grand Banks of Newfoundland during the latter part of the voyage.

The Dutch, in their American colonization, followed the northern route, while the English split, some going partly on the northern track and some, with the Latins, on the southern track. Nantucket Island was probably the dividing line at the American receiving end of these colonists. To the north of Nantucket were the northern route colonists, and in the Carolinas, Georgia and Virginia, those who had followed the route via Equatorial waters. A difference of a hundred miles or so in final destination meant a length of passage which was about 3,000 miles longer by the southern route; who knows whether the difference in ancestry and the inherent characteristics of those ancestors which made them choose one route rather than the other did not finally cause the establishment of two peoples with different values. This in turn led to the American Civil War and to the difference in attitude of north and south which has existed even until today.

As previously mentioned, the importance of the Gulf Stream during war was realized by Benjamin Franklin, and during the War of Independence this knowledge was suppressed because it might be of value to the British Navy. The U.S. Coast and Geodetic Survey realized the value of a thorough knowledge of the fast current which flowed down their eastern seaboard and for many years made detailed measurements of the current strength, both at the surface, and deep under water where its flow characteristics are interesting, even if not important to surface shipping.

The ocean currents maintain their significance to naval history to this day. Submarines are now able to remain submerged for long periods. In the course of their wandering around the oceans they may well at times like to take a silent ride away from spying enemies; this can be done provided there is a good knowledge of currents. Not only is it necessary to study surface streams, but also those deeper currents which are silently stirring the hidden waters of the sea. At the surface the old Gulf of Mexico, scene of so much crooked dealing, ambition and bloodshed, has become a fascinating playground for the modern generation of explorers, who armed with the aqualung can swim down in the transparent sea to search for old wrecks from the sailing days with all the possible excitement of the discovery of bronze cannons and even gold doubloons.

12

POLITICAL ASPECTS

Who owns the Gulf Stream? This may sound a stupid question to ask, but it is one that is today worrying the international lawyers, providing them with a never-ending subject to discuss, and occupying the time of government officials the world over. Not that the problem is specifically mentioned as the Gulf Stream; the present argument concerns the mineral wealth of the oceans, and since the oceans cover two-thirds of the earth's surface there is a large amount of territory involved.

An international convention of 1958 determined that the mineral wealth beneath the sea-bed of the continental shelves should belong to the nation adjacent to any particular piece of shelf, and this ruling has led to satisfactory development of oil and gas projects in many offshore areas of the world. The 1958 agreement was vague in the definition of the 'edge' of the continental shelf, and contained a clause which allowed adjacent ownership to extend to deeper water provided exploitation of minerals was feasible. This has caused some experts to claim rights on the continental slopes, and some would even wish to extend jurisdiction along the three-mile deep ocean floor. However, another school of thought would like to see the sea-bed under the deep oceans preserved internationally as a common heritage of mankind. The argument against international ownership rests on the undoubted fact that it would be difficult for any would-be developer of ocean-bed minerals to obtain a permit from a committee consisting of all the members of the United Nations. On the other hand, international agencies such as those concerned with health and meteorology do operate fairly efficiently, and an international leasing agency for sea-bed minerals would be feasible. An international régime would share any royalties from mineral development with all nations, and it might make a much stronger United Nations, less dependent than it is now on individual national subscriptions. It would also extend the traditional idea of freedom of the high seas, rather than restricting what goes on in the oceans.

Many nations are already forbidding oceanographic work on their adjacent continental shelves on the grounds that scientists are stealing knowledge which may lead to mineral exploitation. An international ownership of sea-bed resources would allow unrestricted scientific research and it might provide the beginnings of a future 'one world'.

Claims to the mineral resources of the ocean floor have been broadened in some instances to include fishing rights. Some South American countries, for example, have extended restrictions on fishing from the traditional three-mile limit to distances of hundreds of miles in order to include the rich areas of fish growth which exist because of the upwelling waters of the Humboldt Current. Although there is no claim concerning the current itself, there would be a great outcry if the current pattern was changed and the rich upwelling area was displaced to be opposite another country's coastline. Diverting the ocean currents is no mere dream. Modern civil engineering is now capable of making changes to the face of the earth which could have far-reaching effects on the ecological system and the climate, while nuclear explosions can be detonated to give disturbances as large as the most impressive natural eruptions.

The effects of the Gulf Stream have been known for hundreds of years, and it is not surprising that serious suggestions have been made for altering the system that Nature has provided. During the American War of Independence that wily inventor Benjamin Franklin (who had made a special study of the Gulf Stream) proposed that consideration should be given to changing the course of the current, in order to plunge Great Britain into a new Ice Age. No doubt the technical details of altering the Gulf Stream proved too difficult at that time, but what was impossible in the days of sail did not appear an insuperable task to Mr. C. L. Riker in the early part of this century.

Carroll Livingston Riker was born on Staten Island, New York in 1853 and lived a life full of new ideas and inventions, mainly concerned with the sea. He may well be considered as one of the pioneers of a most important subject of today—ocean engineering. In his youth he became interested in currents and wave action in the sea and in rivers, and at the age of seventeen designed the hull of the steamboat *Charleston.*

This interest in the oceans was followed throughout his career as mechanical engineer, inventor, farmer and manufacturer. In 1882 Riker played an active part in setting up the first American factory for the manufacture of unfermented grape juice. Another first was the equipping of the pioneer refrigerating steamship for carrying perishable goods. He had already, at the age of twenty-one, designed the first refrigerating warehouse in New York. He played his part in the Spanish-American war of 1898 by designing a new type of torpedo, together with an ingenious tactical plan of floating a string of these torpedoes on to enemy vessels, or an alternative scheme of drawing a line of torpedoes round the enemy fleet.

One of Riker's large engineering projects was the design of the most powerful dredge at that time; this was used to fill the Potomac Flats below Long Bridge at Washington D.C. In 1907, when the government was making plans for the Panama Canal, Riker's experience was enlisted by General G. W. Goethals who was in charge of the project. A lifetime of experience in large-scale earth moving, and in matters pertaining to the oceans and to rivers, gave Riker the idea for what turned out to be a non-starter, but which might have been his greatest civil engineering project of all. This was the scheme to control the Gulf Stream, and it was no vague idea, but rather a detailed plan backed by well thought out methods of attaining the desired objective, and by proposed experiments to find necesary basic information.

In January 1913 Riker presented his *Conspectus of Power and Control of the Gulf Stream* to the President and the Congress of the United States and also 'for the officials of Interested Foreign Countries and others'.*

The object of the Gulf Stream Control was to stop the cold Labrador Current from the north intermingling with the warm waters of the Gulf Stream. The benefits accruing from the project were firstly the elimination of fogs which are one of the hazards to transatlantic shipping in the Newfoundland area,

*Riker was very much an internationalist, and would certainly have approved of an international régime for control of mineral resources of the ocean bed. During the 1914–18 War he originated a plan for the neutral control of the seas, outlined in a joint resolution introduced in Congress in February 1915.

secondly the removal of icebergs from the shipping lanes, and thirdly improved climatic conditions in the countries near the Arctic Circle. The two currents were to be separated by a causeway 200 miles long, built out from Newfoundland along the Grand Banks. The water is only 200–250 feet (60–75 metres) deep on the Grand Banks, which consist of sand, gravel and debris rafted from the north by icebergs. The Grand Banks rise out of the deep ocean for 15,000 feet (4,500 metres), and must have appeared to Riker as an almost completed natural causeway. The temptation to add the last 250 feet (75 metres) was too much for one experienced in reclaiming large areas of swamp, or designing canals to cut across the Isthmus of Panama.

The Labrador Current sweeps across the Grand Banks on a 250-mile front, and in Riker's words 'the Gulf Stream receives a staggering blow from which it never recovers. The eddies, cross-currents and revolving motions thus created, principally in and because of the shoal waters, are the beginning of its end.' The volume of the Labrador Current, flowing at one mile per hour, is equivalent to about 50,000 million cubic yards per hour, which is about half the flow of the Gulf Stream through the Florida Straits. Riker calculated that it would require a million tons of coal a minute to warm the Labrador Current from 1.7°C. (35°F.) to 12.8°C. (55°F.). On the other hand the Gulf Stream has enough heat to melt all the Polar ice in forty days during which it would cool from 24°C. (75°F.) to 1.7°C. (35°F.). The object of the plan, then, was to stop the wasteful mixing of the two currents, using the Gulf Stream warmth to heat the Arctic while the cold Labrador Current was diverted to slide *under* the Gulf Stream in the deep water to the east of the Grand Banks. The causeway or jetty, built out along the Grand Banks until deep water was reached, would be used to alter the course of the Labrador Current.

The jetty would need to extend for 200 miles from a point south of the Virgin Rocks, and would be forty miles wide at the coast, tapering to three miles at the edge of the Grand Banks, a total of 1,000 square miles of filling to a depth of 250 feet being required. The elegance of Riker's scheme was in the method of building the causeway; the construction work was to be carried out by the Labrador Current itself. Riker believed that sand was being carried south by the current, and that it was, therefore, only necessary to make the sand deposit itself on the sea-bed in order to build up the causeway. The actual method proposed to achieve this was quite simple:

> Obstructor in the form of a great rope cable, or its equivalent, saturated with asphaltum and weighted with wire or otherwise, giving the requisite specific gravity or weight to cause it to just sink in the ocean and rest lightly on the bottom or ocean bed, and having lesser specific gravity than the sand or other deposit it will thereby be prevented from sinking into or being buried by the deposit by virtue of its greater buoyancy, it being heavier than the water but very much lighter than sand in sea water.

In other words, as the level of the sea-bed rose, the 'Obstructor' would rise with it. The giant rope (or perhaps a suitably loaded pipeline would be an effective alternative) was planned to be anchored against the force of the current from the north. If the anchoring proved too difficult, Riker believed that a heavier-than-sand obstructor could be used, being lifted each time it became buried by underrunning the cable. It was estimated that three-quarters of the Labrador Current flow would be stopped in two years. The cost estimate was twenty million dollars, but a much smaller sum than this was requested from Congress so that a feasibility study could be carried out. More information was needed concerning the flow of the Labrador Current and especially the amount of sand that was being transported, and which would be needed to fill in the 1,000 square miles of causeway. The journal, *Scientific American*, offered to furnish a corps of experts to accompany the expedition and many firms signed a petition to Congress on 29 October 1912 requesting financial support for the investigation of the Labrador Current. These firms included submarine constructors such as R. J. Packard Co., Harper and Bros., Harvey Fisk and Co. and many others, indicating that Riker's idea had considerable well-informed support.

The head of the New York Branch of the Hydrographic Office was enthusiastic: 'I am constrained to believe in the feasibility of constructing a jetty or rather, of allowing the currents to construct such breakwater or jetty over the Grand

Banks.' The Maritime Association of the Port of New York petitioned Congress to appropriate the funds necessary for an expedition to determine the feasibility of constructing a jetty, while the New York Board of Trade and Transportation endorsed the Bill for the scientific investigation of possible methods to divert the icebergs from the course of transatlantic steamships The International Mercantile Marine Company '. . . trusted that Riker would be successful in getting Congress to appoint such a Committee or commission and grant it necessary funds'. The draft Bill asking for 100,000 dollars for the investigation covered four pages. It presumably got lost in the turmoil associated with the Great War which began in the following year. There was, of course, considerable opposition to the scheme, both from sceptics who maintained that it would not work, and from vested interests of coast wreckers, cod fisheries and fur trappers. However, as Riker pointed out, there would be a bonus of 1,000 square miles of new territory, especially suitable for wild geese, and hence for sportsmen.

Whether Riker's method of

Unimpeded drops of water, and impeded grains of sand,
Will change the Ocean's bottom and the surface of the land

would really work is doubtful, but his proposed exploratory study of the current was a sensible way of approaching the problem. Looking at the problem today, model studies would probably produce some useful information, and lessons are being learned from the many pipelines carrying oil on and under the sea-bed. Movement of sand certainly takes place; pipelines have been undermined at Kharg Island off Iran, necessitating packing with sandbags, but the reverse effect of lifting a line by sand piling up has not been reported. However, this is almost certainly because the lines are weighted with a concrete wrapping to be heavier than sand. In the North Sea, where the concrete has been stripped by fishing trawlers, there is evidence of sideways movement of the steel pipe; and it is possible that a gradual lift might occur if there was a steady drift of sand against a pipe of the right weight.

The causeway or barrage proposed by Riker would probably divert the Labrador Current, although it is possible that there is enough force in this current to destroy a sand barrier.

Natural forces are very powerful; in some parts of the world channels dredged through sandbanks are continually filled by fresh material, as if Nature was defending what it has decided would be. The Riker causeway would, if it were strong enough, turn the Labrador Current to the east, and then the dense, cold, salty water from the north would undoubtedly sink towards the 15,000 foot ocean deeps bordering the Grand Banks. The warm Gulf Stream could then flow on its eastwards course with minimal contact with the cold current. Riker believed that there would be enough northwards momentum to cause the warm Gulf Stream to flow northwards and divide at the southern tip of Greenland, thus providing two warm streams, on either side of Greenland. However, this is not certain, because of the deflection of the Gulf Stream by the underwater obstacles such as the Grand Banks, and because of the clockwise tendency imposed by Coriolis forces (p. 15). Furthermore, as we now understand, the Gulf Stream can be regarded as merely the outer edge of a general rotation of the surface waters of the North Atlantic. On balance, it is probable that Riker's causeway would bring more warmth to the Arctic and to Greenland and north-west Europe, because there would be more warm water that was not neutralized by the cold Labrador Current. However, the total world heat exchange would have been the same as it is now, the place of supplying the heat would have been moved further north if the Labrador Current was deflected and also some of the cold of the Arctic would have dissipated itself in the deep flowing waters in Equatorial latitudes. Riker certainly proposed the proper course of action, which was to make careful preliminary studies of the various currents and to assess both the practicability of his engineering proposals and the possible climatic effects if the project was achieved.

The effect of a warm current along the west side of Greenland might well be to melt a large part of the Greenland ice-cap, since the prevailing westerly winds would bring heat from the warm current in a similar way to the warm winds which transfer heat from the North Atlantic circulating water from the Equatorial Drift. It is possible that Riker's Gulf Stream Control would produce a much milder climate in Scandinavia and Siberia. However, the Gulf Stream is really only the outer edge of the general clockwise water movement in the North

Atlantic, and a small 200-mile projection placed between it and the Labrador Current would probably not affect the main current pattern, which is controlled by the rotation of the earth and the wind pattern resulting from it.

There is some doubt about the efficiency of Riker's method of causing the 200-mile jetty to build itself up from the sea floor. Another piece of Riker doggerel:

The like of a woman's hairpin and the shoestring of the man,
With the aid of friendly nature, will construct this mighty dam

makes the process sound too easy. Apart from the need to ensure that the transport of sand was adequate to fill the 1,000 square miles of causeway, there would surely be a self-destroying action once the Labrador Current began to be deflected to the east. The stream would then flow along the piled-up sand, and would scour away the material that had already been deposited. There is some evidence that a tongue of land did once extend out from Newfoundland across to the Grand Banks; it has probably been cut away to allow the Labrador Current to have a more direct run to the south. For all the criticisms, Riker's idea was a magnificent concept, although it is the type of project that man should keep for the future when he has learnt how the currents of the ocean are controlled and how they affect the climate. We should follow Riker and spend our efforts on the fact-finding measurements and leave the more glamourous applications to the next century. However, as Riker said, 'The simplicity, cheapness and feasibility of the plan appeal at once to the Scientist, Engineer and the Old Salt.'

Other engineers have applied their inventive minds to warming up the Arctic wildernesses. Anyone condemned to work in Siberia must occasionally get ideas of ameliorating the cold winters. Pyotr Borisov, a Russian engineer, proposed some years ago, that the shallow Bering Straits which separate Russia and Alaska, should be dammed. The scheme was to be more than just a closing of one of the Arctic Ocean outlets since it also entailed a vast battery of pumps which would take large quantities of cold water from the Arctic Ocean. This, it was hoped, would eventually cause a slackening of the cold Labrador Current outlet, with a subsequent northward extension of the warm Gulf Stream waters warming both the Canadian and Russian Arctic shores—with a result in fact very similar to that planned by Riker. As is the case with many grandiose inventions, the detail of the Russian scheme is described before the main principles are established. For, example, to counteract the possible cooling of Alaska and east Siberia by the cold Arctic water being pumped into the Pacific, it was proposed that pumping stations should send warm Pacific water into the Chuckchee and Beaufort Seas.

The Bering Straits are forty-five miles across and about 150 feet deep, so that although it would be a major civil engineering undertaking, it is by no means impossible to construct a dam. With the world population increasing and demanding more space, it may be advisable one day to open up the vast areas of land bordering the Arctic Circle. If the fierceness of the weather could be eased by modifying ocean circulation, the long dark winter nights might be bearable. Many contented citizens live in Fairbanks, Alaska, and with modern house construction and plentiful energy supplies the extremes between summer and winter might prove attractive to many others. There is a danger, already mentioned, of altering the total heat absorbed by the earth if large areas of snow and ice are melted. It is possible that warming the Arctic might be a trigger to set in train a general heating up of the earth as a whole.

Another old engineering project to dam narrow waterways is the German idea of closing the Straits of Gibraltar, and shutting off the Mediterranean. The protagonists of this scheme claim that evaporation in the Mediterranean would cause a lowering of a few feet each year, so that after a few decades there would be a hundred feet or so head of water on the Atlantic side of the Straits. This would enable vast amounts of electrical power to be generated, and must be considered in centuries to come if the nuclear fission method of energy production is not feasible on a large scale. There is plenty of energy for mankind arriving each day from the sun, but it needs large area schemes to harness it. The Mediterranean would probably be rather unpleasant for the first few decades as the water level was falling, and muddy foreshores were exposed with their harvest of filth and garbage deposited by previous generations. However, the mess could be cleaned up and might provide a lesson to encourage man to be more careful and considerate in future.

The Gibraltar Straits dam would be about twenty miles long, but would be in water as much as 1,000 feet (300 metres) deep. Although this depth would bring new problems of construction, it is by no means impossible to devise ways of overcoming the difficulties. There might be some side-effects of the Mediterranean dam, because of the flow of heavy salt water over the sill at the Straits of Gibraltar. This water spreads out into the Atlantic, gradually mixing with the cold current at a depth of about 3,000 feet. This flow from the Mediterranean must have its effect on the North Atlantic water circulation, but it is probably only secondary to the main clockwise motion, and if it was stopped would not alter the Gulf Stream, North Atlantic Drift or the climatic effects in northern Europe caused by the currents in the Atlantic.

Some of the advocates of the Gibraltar dam would like to see the Mediterranean sea-level reduced by evaporation to 300 feet below sea-level in a century. This would provide a large amount of reclaimed land, and would also make it possible to dam the Straits of Messina and the gap between Sicily and Tunisia. This in turn would divide the Mediterranean into two, and would allow for more power production by maintaining different levels in the two halves. Extra power generation would also be obtained from rivers which would be flowing into a much lower sea than at present. These rivers would become much faster flowing, and rapid land erosion would take place. Very considerable model work and theoretical computations are needed, however, before any of these earth-changing schemes are started.

Some engineers prefer to make new waterways, rather than block the narrow gaps that exist in today's geography. The narrow isthmus that connects North and South America is, to some minds, just asking to be breached. The Panama Canal provides a passage from the Pacific to the Atlantic, but not a flow of water; the water for the locks is provided by the lake at the higher level portion of the Canal. A straight, wide cut from Atlantic to Pacific could conceivably be made. Whether it would initiate a flow between the two oceans is not certain. It might provide an outlet for part of the Equatorial Drift water of the Atlantic which flows into the Gulf of Mexico, and thus might ease the outflow through the Florida Straits.

To have any appreciable effect the canal would have to be comparable in size to the forty-mile width of the Gulf Stream at its most constricted point.

These engineering pipe-dreams may never be accomplished, but it must be realized that they could be undertaken and with no more application of money and effort than the space programmes of Russia and the U.S.A. This is why such natural phenomena as the Gulf Stream have political implications. Alterations in the flow of water in the oceans will affect many countries, and in some instances there may be long-term effects which will concern the whole world. Because of this it may be a good thing for the human race if some form of international control is exercised over the deep oceans. This control could begin with the allocation of permits to search for and extract minerals, but it could be extended to international control to ensure that free access was maintained for carrying out oceanographic researches. It could encourage such oceanographic fact-finding as will be needed before any projects are started which could have irreversible consequences, so that in the future man's understanding of the world is so thorough that he will no longer ruin the environment.

13

THE GULF STREAM AND THE FUTURE

The Gulf Stream deserves its popularity with mankind for in general it has a benign influence on man's affairs. Since it is one of the most conspicuous currents of the world, it has attracted the attention of many investigators, and is probably the most studied example of ocean circulation. The swift-flowing portion of the North Atlantic circulation sweeps up the west coast of the U.S.A. and is within easy reach of many of the world's most active oceanographic institutes. The Gulf Stream is still the focus of many oceanographic researches, and as we saw in Chapter 8, new large-scale attacks are being mounted to find out exactly what makes the ocean waters move. During the past few years new techniques of anchoring current meters on the sea-bed have been devised, thus opening up possibilities of making simultaneous observations at many more points than the limited number of oceanographic ships has allowed in the past.

As knowledge of the water circulation of the sea increases, a doubt tends to arise as to whether the Gulf Stream should be considered as a separate entity, or whether it would be better regarded as the outer edge of the general North Atlantic circulation. The 2,000-foot-thick body of warm water which is the Sargasso Sea drifts slowly to the south-west due to the action of wind and the forces consequent on the earth's rotation. This warm water is stopped from reaching the North American coast by the so-called Gulf Stream. The Gulf Stream helps to sweep the Sargasso water north-eastwards, so that the Gulf Stream volume increases from 26 million cubic metres per second at the Florida Straits to 65 million off Cape Hatteras. As the Gulf Stream swings to the east to make its journey across the Atlantic it continues its work of forming a boundary to the warm Sargasso Sea water, separating this warm central North Atlantic water from the cold water to the north. As Stommel says, 'the Gulf Stream is not an ocean river of hot water. The intensity of flow of the Stream, the Stream's direction, and its temperature are not primary climatic factors in determining the climate of Europe; but the rôle which it plays in determining the northern boundaries and average temperature structure of the Sargasso Sea must be of critical climatic importance.'

There is a reason for the current on the west side of the Atlantic being greater than the opposite part of the North Atlantic circulation which flows southwards along the east side. A similar pattern occurs in the Pacific and is due to an augmenting of flow by the Coriolis forces on the west side and a counter-current force on the east side. Theory, then, is in agreement with a fast flow on the North American side of the general Atlantic circulation, so the Gulf Stream is, on this argument, merely a small part of a whole. It is convenient, however, to follow history and to regard it as a current in its own right. The almost jet-like flow at the Florida Straits is a clear-cut phenomenon, and the high speed and warmth of the Gulf Stream do make it a most conspicuous feature to seamen. Although the climatic effects attributed to the Gulf Stream are mainly due to the transport of heat by winds from the warm central Atlantic water which is bounded by the Gulf Stream, there must be some benefit to north-west Europe from the currents that continue from the Gulf Stream to the north of Scandinavia. The warm water is presumably largely responsible for keeping the Arctic ice much farther to the north than it is on the western side of the Atlantic. At least the existence of a friendly current is admitted by some, since there is a group of islands called the Gulf Stream Islands off the northern part of Novaya Zemlya—in Arctic Russia.

The mighty river in the ocean of Maury's days has been relegated to being a boundary of a much more general water circulation now that science has learned more about the behaviour of a fluid on a rotating world, and the debunking of the Gulf Stream will continue. Some years ago Professor Iselin of the Woods Hole Oceanographic Institution suggested that the fact that the fast-flowing Stream contained the warm Sargasso Sea waters meant that warmth was being held back from northern lands, not carried to them as had been supposed in the past. If the Gulf Stream could be slowed down the warm water mass could spread farther north and would act as a heat reservoir to warm the westerly winds which blow over northern Europe. The faster the North Atlantic waters circulate the more compact does the system become, so that if a warmer climate is desired for Scandinavia, the Gulf Stream must be slowed down.

The first half of this century was probably a time when the northern hemisphere was warming up. Climatologists delve into old records which are not necessarily direct measurements of temperature or snowfall, but circumstantial evidence—such as the fact that ships carrying coal from the Spitsbergen mines could operate without ice-breakers for only ninety-five days in the year in 1900, whereas the figure had doubled by 1930. There is evidence that pack-ice in the Russian Arctic has decreased from 1893 when Nansen measured the thickness, to the late 1930s when Russian observations showed much thinner ice-sheets. Signs of a thaw over the same period are provided by the retreat of glaciers in Europe and in Alaska. The most recent observations, however, suggest that we have seen the best of the weather for this century, and that the ice is creeping forward once more. It might, therefore, be a good time to be thinking of a gentle warm-up for the North Atlantic. A change of 5–10°C. (9–18°F.) either way could plunge the earth either into a real Ice Age or into an unbearably hot climate. Since 1940 there have been five winters where the temperature averaged below freezing, whereas no such cold times occurred between 1896 and 1939. This seems to account for the fact that older people nowadays speak of the glorious summers of their young days. To make matters worse the winter of 1962–3 was the worst since 1740. A new Ice Age will mean misery for Northern Europe, Siberia and America, with desolation following as the thick ice moves slowly down to southern England. Although the weather is affected by the airstreams which bring warmth from tropical to northern regions, and which in turn are affected by the distribution of warm water by currents, meteorologists and oceanographers are still far from understanding the exact mechanisms of climatic change. The long-range weather forecasts, which have been getting more reliable during the past four years, are based mainly on comparing trends in particular indicators, such as area of snow cover, or temperature or rainfall, with patterns in past years. Unfortunately, although a cold winter is sometimes the precursor of a good summer, there are cases in the past when this was not so. Therefore, long-range forecasting is mainly a statistical process, rather than a calculation of how a future weather pattern will develop from the set of conditions existing today. The difficulty is that there are so many variables that may be important, and a great deal more effort must be put into planning experiments which will decide on alternative theories of the behaviour of air and water on a circulating earth, which is bathed in sunshine except where there is cloud cover. Apart from such experiments, there is a need for collection of regular observations at sea, such as are provided at the present time by the small number of weather ships. Modern development of large buoys which can remain for years transmitting regular observations to a central headquarters will enable us to tackle these jobs in the future. Meanwhile, the more indicators that can be found to correlate with our local weather, the better our statistical forecasts will become. One such indicator which is proving to be useful for the British Isles is the temperature of the water over the Grand Banks of Newfoundland. When this patch of water is cold a high pressure develops in the atmosphere to the north-west of Britain giving warm weather; on the other hand if the sea off Newfoundland becomes warm it means a wet summer for the British Isles. Presumably the sea temperature in this critical area is dependent on the interaction of the Gulf Stream with the cold Labrador Current. If only Mr. Riker had finished his causeway we might have been able to exercise some form of control!

Another bizarre scheme for the Gulf Stream, which would make for a milder climate in the north, is the idea that the Stream should be put to work turning giant turbines situated in the Florida Straits. The turbines would produce vast quantities of electric power, and by removing energy from the fast-flowing current, the Gulf Stream system would be slowed down, resulting, as we have seen, in a spreading of the warm tropical water and therefore providing a larger source of heat for the winds to carry over Britain and Scandinavia.

Some wishful thinking was expressed in a 1923 book by the German author, J. E. Kiesel, who appears to have hated England, and expunged this hatred from his system by a fictitious work on a seven-year war waged by U.S.A. against Europe.

The German hero devised a plan to divert the Gulf Stream, and sold this information to the Americans. The plan seems to have been a modification of Riker's causeway, since an obstruction east of Cape Hatteras was supposed to make the Gulf Stream continue to the north instead of swinging east to Europe. A secondary civil engineering project of some dimension was to pierce the islands of Novaya Zemlya in order to increase the circulation in the Arctic Ocean. Modern knowledge suggests that the scheme would not have worked—in fact the diversion might have improved the European climate. However, the supporting terrors quoted are interesting. The great thickness of ice which formed over Europe was to shift the earth's axis, while a flying submarine was to be employed, presumably to cover the ice like a hovercraft.

The dam to deflect the Gulf Stream was to be formed by dropping cubes of concrete on to the sea bottom. In the book it was the continuous snow in Scotland and mutiny in the Royal Navy which finally brought peace. The Gulf Stream obligingly returned to its former course, as presumably it does after every Ice Age. The reason for the warming up and return to normal was because the anti-British German had a change of heart and finally preached universal love. This is one of the few documented explanations of why Ice Ages cease!

It is surprising that with all our expertise in nuclear warfare and in space travel, we have not found out why the Ice Ages arrived in the past. These periods were quite catastrophic to large parts of the earth and its inhabitants, and from what knowledge we have gathered, it appears probable that quite small beginnings in a weather cycle build up to a general cooling down or heating up which must run its course once it has been started. If there is to be a permanent succession of generations for tens of thousands of years it may be necessary to be able to avert the next incipient Ice Age.

Ewing and Donn (of the Lamont Geological Observatory) suggested some years ago that it is the oceans rather than variations of the sun's heat that are the cause of Ice Ages. The present susceptibility of the earth to enter a cold cycle is due to the particular arrangement of land and sea on the earth's surface, which has not been the same throughout the life of the earth on account of continental drift. A few hundred million years ago most of the land was concentrated in one unit, and one ocean covered the remainder of the globe. The ocean currents, the forerunners of the Gulf Stream, which we know must have existed as a consequence of the rotation of the earth, were sufficient to stir up the Polar waters and ensure that a large ice-cap did not form. In our time the Arctic Ocean is almost landlocked, and the warm Equatorial water cannot mix with the Polar ice and snow to melt it. At the South Pole the large land mass of Antarctica forms a depository for snow and ice, again protecting it against the warm currents.

Perhaps a regulation of earth temperature may be found which can affect the natural forces. For example, if the earth begins to heat up, due to increased emission from the sun, a shield might be produced by putting more dust particles in the atmosphere. Or the amount of reflected sunshine might be increased by enlarging the Polar ice-caps. It might not be all that difficult to extend the Antarctic ice. The thickness of the ice-cap around the South Pole is such that the pressure due to the weight of about two miles' thickness of ice is almost causing the ice to melt where it is in contact with its rock base. Melting would decrease friction, and a whole mass of ice might slide off into the sea, doubling in a short time the white area of the southern hemisphere, since the land left behind by the ice-cap would quickly become covered with snow. Melting might be encouraged by suitably placed nuclear devices providing the small amount of extra heat needed.

The artificial increase in Polar whiteness could be used to plunge us into another Ice Age, rather than being used as an antidote to an overheating earth. However, there is a possible way of counteracting the cooling down; this would be accomplished by sprinkling ash on the snow and ice and therefore ensuring that more of the sun's heat was absorbed. This process might be most effective in the north Polar regions where the snow and ice is thinner than in the south, since extra heat would melt the ice and remove the white reflecting surface altogether. If an ice-cap grows very large it may eventually defeat its propensity to cool the earth, because the sea will become cooler around it, and there will be a tendency to less evaporation, and consequently smaller snowfalls will occur. The large size will mean that inland parts of the ice-cap will be far distant from the source of precipitation, and less snow will be available to feed glaciers, which will eventually recede.

One can dream up many ways of tinkering with the climate of the earth, even using the meagre knowledge we have at present. No one is capable of calculating whether any of these fanciful schemes would in fact work, or on how large a scale they would need to be applied. We do not know how much built-in stability our earth's atmosphere and oceans possess. This is why forward-thinking scientists are pushing for international effort for experiments such as the Mid-Ocean Dynamics Experiments, and for more observations of the atmosphere and of the ocean surface from satellites and of the body of the ocean by means of instruments suspended from buoys or tethered to the ocean floor.

There are many writers today who are preaching Doomsday for the human animal on earth, pointing out that excessive waste, greedy use of raw materials, and pollution of air and water will inexorably result in a dying, dirty world. Even heat has become an offender in the eyes of some ecologists, because it causes the wrong things to grow too fast, so that fresh cool streams become filthy weed-infested sewers with no oxygen to support fish life. The oceans contain 300 million cubic miles of water, and the great currents like the Gulf Stream have a flow which exceeds all the rivers of the world added together, so that it is conceivable that the heat produced on land is acceptable provided it can be transferred to the sea and suitably mixed with it. This is why a thorough knowledge of currents is vital to our present world ecology (see Chapter 3).

The pollution problem has been creeping up on us since the Industrial Revolution and has been accelerated by the rapid strides made in medicine during this century. The main reason for pollution today is the rapid growth in population. It is curious that most of those who preach conservation push the population increase out of their minds, and spend their efforts tackling the consequences rather than the cause. If world population is to double by the end of the century, and if our educational system is to teach people that a better life entails more material comforts, the fight against pollution will be impossible to win. The prime effort should be aimed at limiting the population.

While the human race is busy stabilizing its population, it will be advisable to spend a few decades cleaning up the accumulated refuse of the past. The methods of re-establishing clean water in Lake Erie or in freshening the Rhine are well understood, and are in fact in action already, but they are very expensive. There is only one ultimate source of finance for cleaning the atmosphere, the rivers or the sea, and that is from the pockets of us all. A short-sighted view may naïvely suggest that 'industry' should pay for cleaning up, but a moment's thought will serve to realize that the bill for money spent on cleaner production will of necesssity be passed down the line to the consumer. If people who stand to benefit from holiday-makers do not like raw sewage piped into the sea, they must be prepared to pay higher rates for a modern sewage plant. If the inhabitants of Los Angeles desire smog-free air they must fit after-burners to their motor-cars.

If the clean-up is to be made now it will have to be carried out at the cost of increasing our standard of living. What is even more apparent, and perhaps not acceptable to their people, is that the less developed countries will need to curb their aspirations to a new world full of technological riches. The alternative for these new contenders for the affluent age is to pay no heed to conservation, and to follow the early industrialized countries into the uncontrolled mess of dirty air and water. Enough experience has been gained during the last few decades to show what chaos can result from uncontrolled development of natural resources; but, on the other hand modern engineering skill and experience can provide what humans want in the way of minerals, and yet enable the work to be carried out safely and tidily. A sensible contemporary slogan is that good engineering is good conservation—a well designed project not only includes the avoidance of eyesores and dirt-making, but also takes care of the things that might possibly go wrong. Accidents are expensive to industry, and engineers build into their plan safeguards which take care of the frailty of human endeavour.

If it has appeared that conservation and the effect of increase of world population—'popullution' as some have dubbed it—have strayed into this discussion of the Gulf Stream, there is a good reason. The Gulf Stream is the best known and the most carefully researched current that flows into the oceans. This is because it is one of the largest and fastest of the water circulations of the seas, and because it has always played an important part in the lives of the people who live on the land which borders the North Atlantic. Because of its size and its geographical location the Gulf Stream is the most suitable world current for the investigation of the fundamentals of water circulation in the ocean. Therefore, this account of the Gulf Stream is an unfinished report, and it could well be that the claim to fame of the Gulf Stream to future generations may be based on the understanding of natural processes gained from the investigations being planned for the 1970 Decade of Oceanography.

BIBLIOGRAPHY

*DEACON, G.E.R., and CURRIE, R., *The Oceans.* Paul Hamlyn, London, 1960.

GASKELL, T.F., *World Beneath the Oceans.* Doubleday, New York, 1965.

*GROEN, P., *The Waters of the Sea.* Van Nostrand Reinhold, New York, 1967.

The Story of the Gulf Stream. Van Nostrand, London, 1967.

HEYERDAHL, THOR, *Kon-Tiki Expedition.* Rand McNally, New York, 1968.

*LOFTAS, T., *The Last Resource.* Regnery, Chicago, 1970.

MAURY, H. MATTHEW FONTAINE, *Physical Geography of the Sea.* Harvard University Press, Cambridge, Mass., 1963.

*MILLER, R.C., *The Sea.* Random House, 1966.

*PRAEGER, R.L., *The Way That I Went.* Hodges, Figgis & Co., 1909.

RENNELL, JAMES, *An Investigation of the Currents of the Atlantic Oceans* etc. ed. J. Purdy. London, 1832.

RITCHIE, G.S., *The Admiralty Chart.* American Elsevier, New York, 1967.

*STOMMEL, H., *The Gulf Stream.* University of California Press, Berkeley, 1965.

*Titles marked with an asterisk denote those books which the author feels are particularly useful and which have been of great value to him in preparing this work.

INDEX